Macrocycle-Functionalised Nanosensors

Macrocycle-Functionalised Nanosensors

Weng-I Katherine Chio
Tung-Chun Lee

JENNY STANFORD
PUBLISHING

Published by

Jenny Stanford Publishing Pte. Ltd.
101 Thomson Road
#06-01, United Square
Singapore 307591

Email: editorial@jennystanford.com
Web: www.jennystanford.com

British Library Cataloguing-in-Publication Data
A catalogue record for this book is available from the British Library.

Macrocycle-Functionalised Nanosensors

ISBN 978-981-5129-33-5 (Hardcover)
ISBN 978-1-003-51057-4 (eBook)

Contents

Preface

In the realm of supramolecular chemistry, molecular containers hold great promise for innovation in trace chemical detection. Over the past few decades, the unique physicochemical properties of cucurbiturils have inspired the imagination of scientists, weaving their complex structures into a tapestry of discovery that has not only reshaped our understanding of host–guest complexations but also broadened the horizons of their practical applications. Nanosensors, enabled by ingenious molecular and nanoscale design, have become indispensable tools in addressing some of the most pressing challenges facing our world today, from rapid healthcare diagnostics to on-site environmental monitoring and beyond.

Our goal is to provide the readers with the knowledge, insights, and practical skills required to navigate this exciting domain. We begin with the fundamentals, ensuring a strong foundation in host–guest chemistry, synthesis of nanoparticles, and optical spectroscopy. From there, we journey through the intricacies of macrocycle functionalisation, exploring how these molecular architectures enable us to fabricate nanosensors with high sensitivity and selectivity.

We would like to thank the staff at Jenny Stanford for the kind assistance and editing of the book.

Weng-I Katherine Chio
Tung-Chun Lee
March 2024

Acknowledgements

WIKC is grateful for the support from The Science and Technology Development Fund, Macau SAR under Funding Scheme for Postdoctoral Researchers of Higher Education Institutions 2021 (File no. 0028/2021/APD).

TCL is grateful for the support from the Research Project Grant (RPG-2016-393) funded by the Leverhulme Trust, and the Institute for Materials Discovery, University College London.

Chapter 1

Introduction

This chapter is designed to introduce the basic concepts and fundamental knowledge of supramolecular chemistry, plasmonic nanoparticles, macrocycle-nanoparticle composites, and their sensing applications to prepare the readers for the following chapters.

1.1 Supramolecular Chemistry

Supramolecular chemistry is a fast-growing and interdisciplinary field sparking great interest across different disciplines including chemistry, biology, nanotechnology, and materials science. Supramolecular chemistry describes the non-covalent interactions such as hydrogen bonds, hydrophobic forces, van der Waals forces, π-π stacking, and metal-ligand coordination between two or more molecules or ions to form self-assembled structural complexes and molecular machines mimicking complex biological systems. The dynamic nature of the non-covalent interactions allows the assembly, disassembly, and re-assembly of the molecular architectures to occur in a reversible manner. The strength of the interactions plays an important role in the stability and length scale of the resultant supramolecular assemblies while the directionality of these interactions determines the type of self-assembled structures. Although hydrogen bonds, π-π stacking,

Macrocycle-Functionalised Nanosensors
Weng-I Katherine Chio and Tung-Chun Lee

ISBN 978-981-5129-33-5 (Hardcover), 978-1-003-51057-4 (eBook)
www.jennystanford.com

and metal-ligand coordination are superior to other interactions, a combination of different non-covalent interactions is often employed for higher association constant and specificity, leading to the molecular recognition properties exhibited in macrocycles. The supramolecular motifs can be tailored to recognise and bind to specific molecules by careful design of their topologies and functionalities. Thus, the structures and properties of macrocycles can be modulated by chemical triggers such as competitive guests and inhibitors, as well as external stimuli, such as light, pH, and temperature, for their wide-range applications.

While numerous macrocyclic hosts have been reported since the blooming of supramolecular chemistry, including, but not limited to, crown ethers, molecular tweezers, cyclophanes, cavitands, porphyrins, catenanes, and rotaxanes [1], this textbook features the host–guest chemistry of cavitands (Fig. 1.1) such as calixarenes, cyclodextrins (CDs), cucurbiturils (CBs), and pillararenes (see Chapter 2).

1.2 Plasmonic Nanoparticles

Nanomaterials generally refer to materials with at least one dimension less than 100 nm, ranging from zero-dimensional nanoparticles (NPs) to two-dimensional thin films. The small size of NPs implies that they exhibit a huge surface-to-volume ratio and thus surface effects become dominant for this class of nanomaterials. They offer distinct and significantly improved optical, chemical, and biological properties compared to their bulk counterparts. In addition, NPs often have a discontinuous electronic band structure which results in unique size-related electronic, magnetic, and optical properties (quantum size effect) and have stimulated intensive interest in nanotechnology.

The very first application of plasmonic NPs can be dated back to the 4th century when the Lycurgus Cup (Fig. 1.2), the most well-known example of ancient artefacts composed of noble metal NPs, was manufactured by Roman craftsmen (now in the British Museum). Interestingly, the Lycurgus Cup resembles jade with an opaque yellowish-green colour under reflected light but resembles ruby with a translucent purplish-red colour under transmitted illumination. The dichroic nature of the Lycurgus Cup is due to the

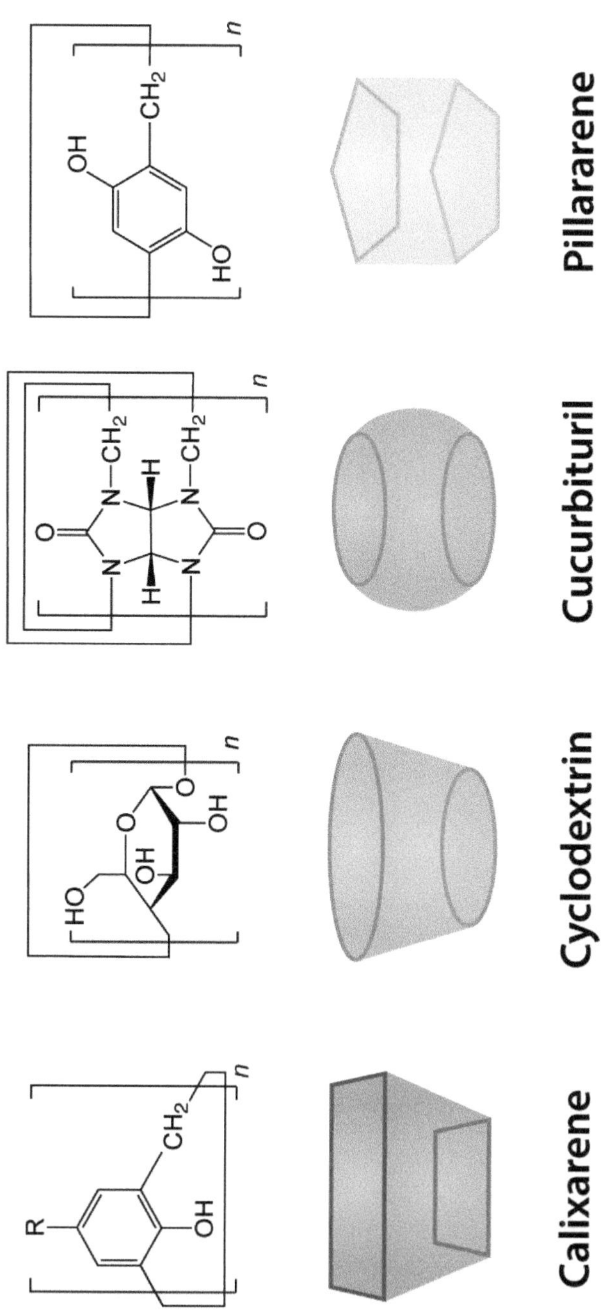

Figure 1.1 Chemical structures and schematic illustration of different cavitands.

presence of composites of Au and Ag NPs at a ratio of 3:7. The distinct colour appeared on the plasmonic NPs results from a photophysical process known as surface plasmon resonance (see Section 4.7.1).

Figure 1.2 The Lycurgus Cup (a) when illuminated from outside and (b) when illuminated from inside. Copyright The Trustees of British Museum.

Although Francisci Antonii, a philosopher and medical doctor, published the first book on the concept of colloidal gold in 1618 [2], no scientific evidence had been reported until Michael Faraday produced a deep red solution of Au NPs by reducing an aqueous solution of chloroaurate ($AuCl_4^-$) using phosphorus in a two-phase CS_2-water mixture in 1857. He also investigated the optical properties of thin films prepared from dried colloidal solutions and observed reversible colour changes from bluish-purple to green upon mechanical compression [3].

Since then, a wide range of methods, which can be categorized into top-down and bottom-up approaches, have been developed for the preparation of NPs. Top-down approaches refer to the fabrication of NPs via gradual size reduction from bulk materials while bottom-up approaches refer to the synthesis of NPs from atoms, molecules, and clusters. Attrition milling and lithography are typical examples of top-down approaches, which are relatively cheap and suitable for mass production, although it is often challenging to process the materials at the nanoscale with good homogeneity. On the other hand, colloidal self-assembly and chemical vapour deposition are two examples of bottom-up approaches which are popular for their relatively narrow size distribution and high uniformity.

Nowadays, plasmonic NPs with different shapes, ranging from spheres, cubes, rods, and triangular/hexagonal plates to tetrahedrons, octahedrons, bipyramids, and stars, can be synthesised by a great variety of methods and some of them have even become commercially available. This textbook specifically focuses on colloidal self-assembly as it includes the most commonly used synthetic routes for plasmonic NPs with control over size and shape which is desirable for sensing applications (see Section 3.2).

In addition to synthesis and fabrication, the ability to visualise NPs plays an important role in nanotechnology. Materials characterisation techniques including optical spectroscopy and electron microscopy allow us to visualise NPs (see Section 3.5) and thus establish their structure-property relationship for potential applications in photonics and catalysis.

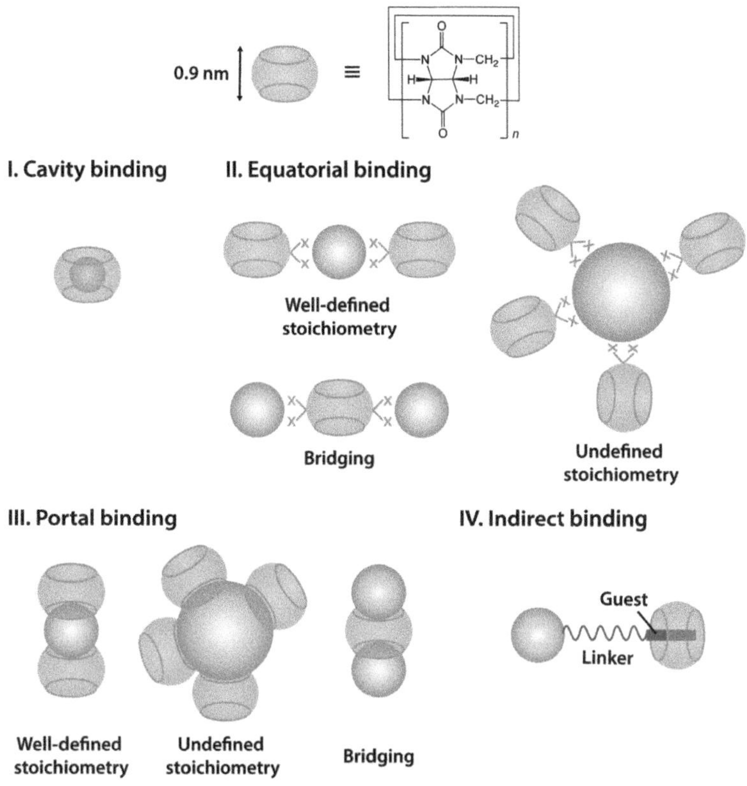

Figure 1.3 Schematic illustration of different binding modes of macrocycle-NP composites.

1.3 Binding Modes of Macrocycle-Nanoparticle Composites

Hierarchical aqueous self-assembly of plasmonic NPs triggered by macrocycles provides a facile method to fabricate nanocomposites for a wide range of applications such as plasmonic sensing. In theory, there are three possible ways (Type I–III) to directly bind macrocycles with well-defined molecular structures such as CBs to plasmonic nanoparticles such as Au NPs while there is also a way (Type IV) to indirectly attach the macrocycles onto NPs (Fig. 1.3).

1.3.1 Type I Cavity Binding

Cavity binding refers to the encapsulation of a bare metal NP within the cavity of a macrocycle. For instance, the highly rigid structure and extremely low polarisability of CBs create an unprecedented nano-environment for various interesting properties of the encapsulated moieties to arise, such as the strong fluorescence enhancement and quenching of dye molecules upon CB encapsulation [4]. Moreover, the two symmetric portals of CB allow access of small molecules to the encapsulated species in a specific manner. For Type I binding, this control over access to an encapsulated metal NP serves as a novel strategy in constructing hierarchical functional assemblies and nano-reactor systems.

Although the possibility of forming a Type I CB7-NP composite was first proposed in 2007 [5], it remains improbable, or at best debatable, that a metal cluster can be encapsulated within the CB cavity via Type I binding based on the current reports in the literature. Indeed, it is extremely challenging to encapsulate a bare metal NP inside the CB cavity, even for the largest common CB homologue (CB8) according to both experiments and computer simulations. While CB8 can accommodate 21–47 gold atoms in its cavity, it is still less than the magic number (number of atoms = 55) for noble metal clusters [6]. The metal clusters at this size range are often difficult to synthesise in bare form due to their exceptionally large surface-to-volume ratio and the quantum mechanical effects associated with the molecular orbitals of a specific cluster [7]. As a result, they are usually prepared in organic solvents and stabilised by bulky and strong ligands, such as aryl phosphines. Bare or semi-bare metal clusters, which are highly unstable in solution and believed to only exist in the prenucleation stage of NP synthesis, are required for forming Type I macrocycle-NP composites. The metal clusters could be generated transiently by either in situ reduction of a metal precursor or by ligand displacement reaction from ligand-stabilised metal clusters. Moreover, significant interactions between the CB portals and metal ions/clusters (either neutral or positively charged) represent another barrier for the formation of Type I CB-NP composites. For instance, the Ag^+ ion was found to be more stable at the CB portals than inside its cavity via DFT-based computational calculations. Unfortunately, characterisation is also not a trivial task as transmission electron microscopy (TEM), the conventional technique for visualising metal NPs, becomes ineffective for characterising metal clusters with <20 atoms. Nonetheless, mass spectrometry could be used as an ultrasensitive and accurate technique for characterising metal clusters and their complexes with CBs and showed the preference for binding between the clusters and the CB portal.

1.3.2 Type II Equatorial Binding

Equatorial binding consists of binding interactions or bond formations between metal NPs and the functionalities attached to the tertiary carbons on the glycoluril units of CBs, resulting in the major C_n axis of CBs lying parallel to the surface of NPs. While the functionalisation procedures for CBs reported in 2003 offer new opportunities for attaching different functionalities to the equatorial positions of CBs and incorporating them into functional networks [8, 9], the procedures cannot be easily reproduced. Moreover, a mixture of products containing partially functionalised CBs is often formed in the functionalisation reaction, together with other by-products that cannot be readily purified. In 2011, the first Type II CB-NP composite between thiolated-CB6 and CdTe quantum dots (QD) was reported [10]. The thiolated-CB6 bound to the QD surface have retained their host–guest binding capability, which was then utilised in the complexation of nitrobenzene amine guests onto the QD surface. The photo-induced electron transfer properties of the resulting Type II CB-QD demonstrate potential applications in molecular-recognition biosensors, light-emitting diodes, and photovoltaic cells.

1.3.3 Type III Portal Binding

Portal binding builds on the supramolecular interactions between the electron-rich carbonyl portals of the macrocycles and the metal

NP surface. In 2008, the presence of supramolecular interactions between the CB portals and a flat Au surface was determined using high-resolution X-ray photoelectron spectroscopy (XPS) and Fourier transform infrared reflection absorption spectroscopy (FT-IRRAS) [11]. The surface-bound CBs generally retain their ability to form host–guest inclusion complexes. Moreover, the surface-bound CBs were shown to supramolecularly immobilise a guest-functionalised yellow fluorescent protein onto a flat Au surface in a reversible manner [12]. This strategy was then employed in studying surface-enhanced Raman spectroscopy (SERS) of CBs through the formation of a self-assembled monolayer (SAM) of CBs on a sub-micron structured Au surface (Klarite™) [13]. Meanwhile, the first Type III CB-NP composites based on the self-assembly of CB5 and Au NPs were also reported with the composites forming dynamic aggregates in aqueous media [14].

Soon after, the synthesis of Type III CB6-Pd NPs and their application as a catalyst was also reported, showing reasonably good catalytic activities with enhanced lifetime and recyclability while CB6-Pd NPs remained insoluble in aqueous media [15]. In late 2010, the synthesis of CB7-Ag NPs and CB7-Au NPs was also reported, with CB7-Ag NPs showing potential applications in cancer therapeutics [16, 17]. Unfortunately, the size distribution of the CB7-Au NPs was relatively broad. Subsequently, CB7-Au NPs were used as an optical probe for enzyme-linked immunoassays [18]. In particular, ammonium ions (NH_4^+) generated from the enzymatic reaction of urease were found to affect the degree of aggregation of the CB7-Au NPs dynamic aggregates, which can then be reported by a change in the UV-vis spectra or in the hydrodynamic size using dynamic light scattering (DLS) techniques. Then, CB7-Ag_2S nanocomposites with CB7 as a pre-organisation template were also fabricated [19]. Meanwhile, the supramolecular uptake and release of a phototherapeutic porphyrin dye by CB7-Ag NPs was reported but the Ag NPs within the composites had a very broad size distribution and the fluorescent spectroscopic results were not very convincing [20].

The first systematic study on Type III CB*n*-Ag NPs for all the common homologues (n = 5–8) was reported [21]. However, there is a discrepancy between the reported solution stabilities of CB5-Ag NPs and a later finding that showed CB5-Au NPs and CB5-Ag NPs were stable in solution for three months [14]. This could be due to the low purity of the commercially available CB5 used in the former study. Later, a self-calibrated in situ SERS sensing system utilising Type III CB*n*-Au NPs for molecular recognition was reported while the unique capability of CB*n* in defining sub-nanometre precise hotspots between Au NPs for SERS applications was demonstrated in a separate study [22, 23]. These initial studies formed the basis of a series of research work on translating such systems into (bio) chemical sensors for detecting molecular targets in practical scenarios, which will be further elaborated in Chapters 5–7.

1.3.4 Type IV Indirect Binding

Indirect binding refers to the binding between a macrocycle and suitable guests that are covalently anchored on the metal NP surface via the formation of host–guest inclusion complexes, typically producing the rotaxane or pseudo-rotaxane type of structures on the NP surface. In contrast to Type I–III bindings, Type IV binding does not involve direct interactions between any part of the macrocycle and the surface of metal NP. These types of structures were shown to offer great promise in organic media, whereas building them up in aqueous media is non-trivial and challenging. Since the formation of supramolecular complexes often requires hydrophobic guest moieties extending out from the NP surface in an aqueous environment, it is necessary to carefully address the solubilisation and functionalisation issues of the guests and metal NPs.

In 2010, the first Type IV CB-NP composite based on CB7 and an alkyl-diammonium functionalised Au NP (Au NP-NH_3^+) was reported [24]. The cytotoxicity of Au NP-NH_3^+ was significantly reduced upon threading of CB7 onto the Au NP surface and the composite was then subjected to uptake by cancer cells. Subsequent intracellular triggering of the therapeutic effects of Au NP-NH_3^+ was achieved through administration of a competitive guest (1-adamantylamine) which removed CB7 from the NP surface, causing the endosomal release and concomitant in situ cytotoxicity of the Au NP-NH_3^+. In the following year, a three-component Type IV composite based on CB8, methyl viologen (MV^{2+})-functionalised Au NPs, and a multivalent polymer with pendant 2-naphthol (2-Np) functionalities was reported [25]. In the system, CB8 was used as a supramolecular glue, bringing together the MV^{2+}-functionalised Au NPs and the 2-Np-functionalised polymer, hence forming a self-assembled polymer film interpenetrated by a single layer of close-packed Au NPs within the matrix. As CB8 can simultaneously encapsulate two complementary guests inside its cavity, it has been employed in ligating a wide variety of small molecules and soft materials appended with appropriate guest moieties. One-step fabrication of microcapsules held together by 1:1:1 host–guest complexes of CB8, MV^{2+}, and 2-Np at a liquid–liquid interface was also demonstrated [26].

1.4 Role of Macrocycles in Nanocomposites

The capability of macrocycles to simultaneously stabilise and mediate the self-assembly of NPs to form plasmonic nanojunctions meanwhile encapsulating analyte molecules close to the surface of NPs in an aqueous environment is particularly important for sensing molecular targets with intrinsically low affinities to metal surfaces.

1.4.1 Stabilisation of NPs in Aqueous Environment

Macrocycles such as calixarenes, CDs, and CBs can stabilise metallic NPs upon the formation of macrocycle-NP nanocomposites. For instance, native calixarenes (calix[4]-resorcinarenes), which are organic soluble macrocycles with phenyl rings bridged by methylene linkages (Fig. 1.1), were found to stabilise Au NPs with diameter up to 20 nm in mesitylene, despite the weak affinity between Au and O [27]. The cooperative adsorption of multiple O atoms in the macrocycle passivates the metal surface in the absence of other strongly chemisorptive species. Nevertheless, potential applications of calixarenes-NP nanocomposites in molecular recognition have been greatly limited by their lack of water solubility. In 2005, a mixed self-assembled monolayer approach was developed to synthesise water-soluble calixarene-Au NP nanocomposites with host–guest binding properties [28]. In particular, calixarene-Au NP nanocomposites were prepared via one-step stabilisation and functionalisation of citrate-capped Au NPs with a 2:1 mixture of sulfanylalkyl oligo(ethylene glycol) and thiolated-calix[4]arene in a water/THF solvent system.

Cyclodextrins (CDs), which are a class of water-soluble, naturally occurring cyclic oligosaccharides with ease in functionalisation, are excellent candidates for fabricating macrocycle-NP nanocomposites in aqueous media (Fig. 1.1). Since the native form of CD has weak interactions with metal NP surfaces, it is necessary to functionalise the CD molecules before linking them to the metal NPs, typically via thiolation of primary hydroxyl groups on the minor portal of CDs [29, 30]. On the other side, α-CD-AuNPs can be synthesised by in situ native CD reduction of $HAuCl_4$, forming dynamic nanoaggregates of α-CD-AuNPs in aqueous solutions. During the Au NP synthesis, α-CD acts as a both capping and reducing agent. In particular, the primary hydroxyl groups of α-CD reduce Au^{3+} to metallic Au, with itself being oxidised to carboxylic acid [29]. The negatively charged oxidation product of α-CD provides an efficient capping for the Au NP surface due to the Au-COO^- interaction, thus stabilising Au NPs in an aqueous environment both sterically and electrostatically.

Cucurbiturils (CBs) are a family of water-soluble, rigid, and symmetric macrocycles (Fig. 1.1). Compared to cyclodextrins, they are relatively less soluble in pure water and more difficult to functionalise. The first CB-NP nanocomposites synthesised via aqueous self-assembly were reported in 2010 [14]. For instance, CB-Au NP nanocomposites can be synthesised either via $NaBH_4$ reduction of an aqueous solution of $HAuCl_4$ in the presence of CB*n* (*n* = 5–8) or by simple mixing of CB solution and pre-formed citrate-capped Au NPs [14, 31, 32]. The latter approach is more robust and versatile as the Au NP growth is separated from the CB-mediated self-assembly. The feasibility of forming the CB-Au NP nanocomposites both before and after the reduction of $HAuCl_4$ shows the formation mechanism is thermodynamically driven instead of kinetically trapped. The initially formed Au NPs are stabilised by CBs which prevent them from excessive clustering in the reduction step. The electron density of CB's electron-rich carbonyl portals is polarised towards the Au NP surface upon binding, weakening the C=O bond. The CB-Au NP nanocomposites are stable in an aqueous environment for up to three months with no precipitation, thanks to the presence of Na^+ cations on the vacant carbonyl portals of singly capped CBs which electrostatically stabilise the nanoaggregates [31].

1.4.2 Mediating Self-Assembly of NPs to Form Plasmonic Nanojunctions

Macrocycles can form host–guest complexes via aqueous self-assembly driven by hydrophobic and electrostatic interactions. If both portals of the macrocycle are not sterically obstructed by the guest molecules, the resultant host–guest complexes typically retain the ability to bind to the plasmonic NP surfaces, thus mediating the formation of macrocycle-NP nanocomposites. If one of the portals of the macrocycle is sterically hindered, the aggregation can still be mediated by the empty macrocycles that are in dynamic equilibrium with the host–guest complexes in the system [33, 34]. Precise plasmonic nanojunctions can be formed by simply adding an aqueous solution of the host–guest complexes to a solution of plasmonic NPs, where the macrocycles will rapidly mediate the formation of the nanojunctions (Fig. 1.4). In this case, the interparticle spacing is precisely controlled which is consistent with the portal-to-portal dimension of the macrocycle. This allows strong coupling between localised surface plasmon resonance (LSPR) of adjacent plasmonic NPs, thus enabling strong electric field enhancement within the nanojunctions. The ratio of the doubly capped (bridging) to singly capped (non-bridging) macrocycles depends on the molar ratio between the macrocycle and plasmonic NPs. For instance, increasing the molar ratio of CB:Au NP leads to a decrease in the ratio of doubly capped to singly capped CB, thus decreasing the size of the CB-Au NP nanocomposites. The aqueous self-assembly process of the nanocomposites is mainly driven by the enthalpic gain from the carbonyl-Au interaction and the entropic gain from the release of surrounding water molecules (see Section 3.3 for details about aggregation kinetics). Note that macrocycles with gold-binding moieties only on one of their portals, e.g., thiolated CD [35], or macrocycles with a flexible molecular structure, e.g., thiolated calixarenes [36], cannot mediate the formation of well-defined plasmonic nanojunctions in general.

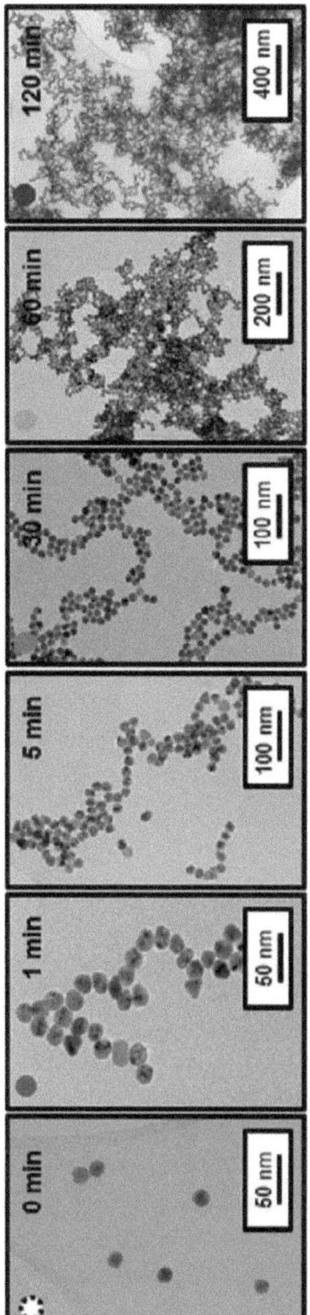

Figure 1.4 TEM images of the CB-Au NP nanocomposites with increasing time [23].

1.4.3 Encapsulating Analyte Molecules Close to the Surface of NPs

Host–guest chemistry of macrocycles enables trapping of analyte molecules within their cavities in aqueous solution. CBs outperform classical macrocyclic hosts such as calixarenes and cyclodextrins for their much stronger binding and higher selectivity towards the shape and charge of guest molecules in aqueous media. Among all, the water-soluble CB homologue, CB7, stands out for its remarkably high binding affinities for cationic and neutral guests in aqueous solution, whilst similar binding affinities can only be achieved for oppositely charged host–guest complexes in other macrocyclic systems [37]. For instance, CB7 has binding affinities of $\sim 10^{15}$–10^{17} M^{-1} with guest molecules such as ferrocene, adamantanes, and diamantanes derivatives, reaching and even exceeding those of biotin-avidin pairs [38]. The ultrahigh binding affinities can be attributed to optimal size, shape, and charge complementarity between the rigid CB cavity and the guest molecules. The ability of macrocycles to trap analyte molecules inside their hydrophobic cavities via host–guest complexation is particularly important for the detection of target analytes with poor affinities to the NP surface, as the encapsulated guest molecules can be positioned in close proximity to the plasmonic hotspots, or even localised at the centre in the case of CB, within the macrocycle-NP nanocomposites.

1.5 Plasmonic Sensing

The distinct colour appeared on the metallic NPs such as Au NPs and Ag NPs is caused by surface plasmon resonance (SPR), which refers to the collective oscillations of free electrons within the NPs when they are polarised by the electromagnetic field of the incident light (see Section 4.7.1). This phenomenon is known as far-field scattering. For instance, the SPR peak of Au NPs and Ag NPs lies within the visible range, at around 520 and 420 nm, respectively. The position and bandwidth of the SPR peak depend weakly on the particle size but strongly on the arrangement, including geometry, interparticle spacing, and degree of aggregation, of the NPs as a result of the plasmonic coupling effect. Functionalised NPs that can aggregate or de-aggregate in response to external stimuli (e.g., light, pH, temperature, and chemical triggers) are often used as sensitive sensors [39]. In particular, plasmonic nanoaggregates are often used as optical sensors for a wide range of applications such as on-site healthcare monitoring, disease diagnostics, environmental monitoring, and homeland security.

There are two distinct forms of SPR, surface plasmon polaritons (SPPs) and LSPR. SPPs are propagating oscillations at the interface between a metal and a dielectric medium with their sizes being much larger than the wavelength of incident light in all directions [40]. SPPs are modulated by the refractive index of the surrounding medium while playing a role in modulating SERS and surface-enhanced fluorescence (SEF) [41]. On the other side, LSPR is a non-propagating, coherent oscillation highly localised to metallic nanostructures (e.g., NPs) with the sizes being less than the wavelength of light. LSPR is strongly dependent on the refractive index of the medium while the local electromagnetic field affects light-matter interaction including scattering, absorption, and fluorescence, thus resulting in plasmonic sensing techniques such as SERS, SEF, and surface-enhanced infrared absorption spectroscopy (SEIAS). Both SPPs and LSPR can be exploited for plasmonic sensing by tailoring the plasmonic nanostructures. In general, plasmonic sensors based on solid substrates support SPP mode whereas those based on colloidal NPs support LSPR mode. The two modes can also be mixed to form SPP/SPP, LSPR/LSPR, and SPP/LSPR couplings [41].

1.5.1 Solid-Based Plasmonic Sensing

Solid-based plasmonic sensors utilising SPP excitation generally adopt the Kretschmann configuration, which can be used to measure the time- and angle-resolved reflectivity SPR response (Fig. 1.5) [42]. In this set-up, the incident light is reflected by a plasmonic metal thin film after passing through a glass prism. When the angle of incidence is greater than the critical angle, total internal reflection occurs, leading to the excitation of an evanescent field on the other side of the prism [40]. When the thickness of the metal film is small enough (~40 nm), the decaying evanescent field can reach the metal-analyte interface, and the resonant condition for SPPs is met for a specific angle of incidence known as the SPR angle. When the spatial modes of the evanescent field and SPPs overlap significantly, the total internal reflection is attenuated, and the incidence light is converted to a SPP [40]. When the analyte molecules bind to the ligands immobilised on the metal film, the SPP peak red-shifts due to the higher refractive index of the molecules with respect to the solvent, typically water. The change in refractive index can be exploited for plasmonic sensing and its sensitivity, S, is given by:

$$S = \frac{d\lambda_p}{dn} \tag{1.1}$$

where λ_p is the wavelength of the plasmon peak and n is the refractive index of the surrounding medium.

The performance of the sensor can then be evaluated by the figure of merit, FOM, which is given by:

$$FOM = \frac{S}{FWHM} \tag{1.2}$$

where $FWHM$ is the full width at half-maximum (i.e., the width of the plasmon peak at half of its maximum amplitude).

The sensitivity can be improved by either overlapping the SPR peak of the metal thin film with the absorption peak of the analyte molecules bound to the substrate or by using a nanostructured film instead of a planar film for a larger surface area and thus stronger local EM field.

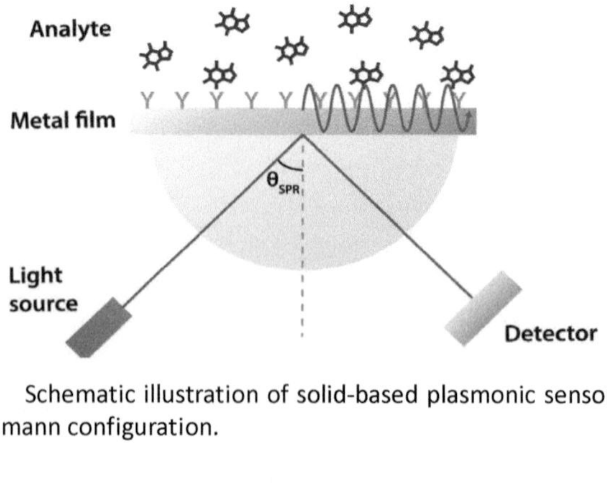

Figure 1.5 Schematic illustration of solid-based plasmonic sensor based on the Kretschmann configuration.

1.5.2 Solution-Based Plasmonic Sensing

Solution-based plasmonic sensors typically utilise colloidal NPs such as Au and Ag which have tunable LSPR peaks in the visible region as aforementioned. When the analyte molecules of different concentrations are added to the colloidal NPs, their interactions result in a shift in the LSPR peak or even a visible colour change due to the change in refractive index. For instance, the colour of the Au NP solution changes from red to purple upon aggregation, providing a facile method to detect the presence of analyte molecules with the naked eye (Fig. 1.6). Colorimetric methods based on functionalised Au NPs are thus widely used for chemical and theranostic sensing (e.g., heavy metal ions, specific prostatic antigen, and human immunoglobulin G), for its biocompatibility, low cytotoxicity, and chemical stability. On the other hand, Ag NPs are often exploited for their stronger EM field enhancement and better plasmonic photothermal properties (see Section 4.6). The sensitivity of Ag NPs is about threefold higher than that of Au NPs of the similar size [43].

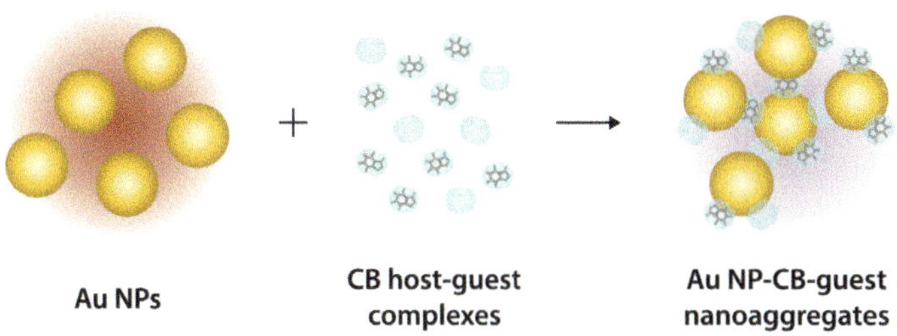

Figure 1.6 Schematic illustration of colorimetric sensors based on colloidal Au NPs.

Apart from being material dependent, LSPR is highly dependent on the geometry of the colloidal NPs, with anisotropic nanostructures having higher sensitivity towards the refractive index compared to their spherical counterparts. The condition for LSPR is met when

$$\varepsilon_1^* = \varepsilon_1\left(\lambda^*\right) = -\chi n^2 \tag{1.3}$$

where ε_1 is the real part of the dielectric constant of the plasmonic NPs (ε_1^* at resonance), λ^* is the wavelength of the LSPR peak, and χ

is the shape factor (a quantity representing the aspect ratio of the plasmonic NPs of which the LSPR mode is based on, e.g., χ = 2 for spheres).

In general, the sensitivity is proportional to the shape factor and is given by [44, 45]:

$$S = \frac{d\lambda^*}{dn} = \frac{\dfrac{d\varepsilon_1{}^*}{dn}}{\left(\dfrac{d\varepsilon_1}{d\lambda}\right)_{\lambda^*}} = \frac{-2\chi n}{\left(\dfrac{d\varepsilon_1}{d\lambda}\right)_{\lambda^*}} \tag{1.4}$$

The larger the shape factor, the larger the aspect ratio and the redder the LSPR peak wavelength. Longer LSPR peak wavelength can have an evanescent field that penetrates deeper into the surrounding media (i.e., increasing the effective sensing volume). Concentrating EM field enhancement around sharp tips or corners is more often used to rationalise near-field effects, e.g., SERS [45]. For example, the sensitivity of Au nanorods is about 3–6 times higher than that of Au nanospheres of similar size [46, 47]. In addition, the sensitivity of plasmonic sensors also depends on the distance between the colloidal NPs, as the EM field strength is highly dependent on the size of the nanogaps.

While solution-based colorimetric sensors provide a facile method for the detection of chemicals and biomarkers, they have relatively lower sensitivities compared to the solid-based sensors with other drawbacks such as false-positive resulted from the background matrix. Other plasmonic sensing techniques based on LSPR, e.g., SERS and SEF, are often used for high-performance sensing platforms (see Sections 4.7 and 4.8).

1.6 Other Sensing Techniques

Apart from plasmonic sensors, a wide variety of sensing techniques is commonly used for the detection of chemicals and biomolecules, including, but not limited to, high-performance liquid chromatography-mass spectrometry (HPLC-MS), real-time polymerase chain reaction (PCR), enzyme-linked immunosorbent assay (ELISA), lateral flow assay (LFA), and electrochemical techniques.

HPLC-MS is an analytical technique that combines separation of mixtures by HPLC and identification by MS. It can be used for quantification of thermally unstable and non-volatile samples such as drugs, vitamins, and proteins [48]. HPLC contains a liquid mobile phase (organic solvents or aqueous solutions) and a stationary phase (e.g., silica) densely packed in a steel column with high pressure and a good flow rate. Different components in the injected compound are separated according to their relative interactions with the two phases, and the output (i.e., separated components) is transferred to MS via an interface (e.g., electrospray ionisation interface and atmospheric pressure chemical ionisation interface). The separated components are ionised, fragmented, and accelerated through a mass analyser in MS and the ions can then be identified by their mass-to-charge (m/z) ratios. HPLC-MS is highly sensitive but it is time-consuming and requires expensive instruments.

Real-time PCR, which is a bioanalytical technique based on the coupling of conventional PCR and fluorescence, is mainly used in disease diagnosis (e.g., SARS-CoV-2), forensic science, and food science. PCR involves adding primers, short DNA fragments complementary to a specific segment of DNA. The four nucleotides and DNA polymerase, an enzyme that can recognise the end of primers, are added to build up the DNA sequence complementary to the original segment. Real-time PCR allows a specific segment of DNA to be duplicated through thermal cycling and quantified by the intensity of the fluorescent signal after each cycle. It is sensitive, selective, and capable of determining different variants of pathogens while having several drawbacks such as being time-consuming and prone to false-positive or negative [49].

ELISA is an immunoassay technique commonly used to detect the presence of antibodies, antigens, proteins, and glycoproteins in biological samples for clinical applications such as diagnosis of HIV infection and pregnancy tests. The measurements are typically carried out in a 96-well plate which acts as the solid phase. The plate is first coated with capture antigen/antibody before the detection antibody/antigen labelled with an enzyme (e.g., horseradish peroxidase and alkaline phosphatase) is added [50]. The unbound species are then removed, followed by the addition of the enzyme's substrate to the plate. The resulting colour change can be measured with a plate reader. ELISA can be classified into four main types, direct, indirect, sandwich, and competitive, depending on the species being coated and detected. They are highly sensitive but expensive, time-consuming, and prone to interferences from the background matrix.

LFA is a paper-based biosensor used to screen for viruses (e.g., SARS-CoV-2 and HIV), toxins, and pollutants in point-of-care diagnosis and environmental monitoring within 5–30 minutes. LFA can be classified into lateral flow immunoassays (LFIA) and nucleic acid lateral flow assay (NALFA). The former uses antibodies as the recognition element while the latter uses nucleic acids [51]. LFIA can be further divided into sandwich type and competitive type based on the detection approach. During the test, a few drops of the sample are added onto the sample pad of the test strip which automatically flows through zones containing different analytes. The sample binds to the capture agent in the conjugate pad and moves to a porous membrane containing a test line (TL) and a control line (CL) where antibodies are immobilised. The test is valid if CL is present and positive if TL is present. LFA is cheap, portable, and user-friendly while having lower sensitivity than PCR and ELISA.

Electrochemical sensors, which utilise electrodes as transducer elements to measure electrochemical reactions between the analyte molecules and the electrode surface, can be subdivided into three different types, amperometric, potentiometric, and conductometric, according to the change in current, potential, or conductivity, respectively. They are commonly used in the detection of environmental pollutants (e.g., pesticides and herbicides) and biomarkers (e.g., glucose and urea), and some are commercially available [52]. Among all, the enzymatic amperometric glucose sensor is a well-known example that measures the interaction with enzymes such as glucose oxidase, with good sensitivity, selectivity, reproducibility, and low cost [53].

References

1. Steed, J. W. and Atwood, J. L. (2009). *Supramolecular Chemistry*, 2nd ed. (John Wiley & Sons, Ltd).

2. Antonii, F. (1618). *Panacea Aurea-Auro Potabile* (Bibliopolio Frobeniano, Hamburg).

3. Faraday, M. (1857). The Bakerian lecture: Experimental relations of gold (and other metals) to light, *J. Philos Trans.*, **147**, pp. 145–181.

4. Hennig, A., Bakirci, H. and Nau, W. M. (2007). Label-free continuous enzyme assays with macrocycle-fluorescent dye complexes, *Nat. Methods*, **4**, pp. 629–632.

5. Corma, A., Garcia, H., Montes-Navajas, P., Primo, A., Calvino, J. J. and Trasobares, S. (2007). Gold nanoparticles in organic capsules: A supramolecular assembly of gold nanoparticles and cucurituril, *Chem. Eur. J.*, **13**, pp. 6359–6364.

6. Schmid, G. (2008). The relevance of shape and size of Au_{55} clusters, *Chem. Soc. Rev.*, **37**, pp. 1909–1930.

7. Hakkinen, H. (2008). Atomic and electronic structure of gold clusters: Understanding flakes, cages and superatoms from simple concepts, *Chem. Soc. Rev.*, **37**, pp. 1847–1859.

8. Jon, S. Y., Selvapalam, N., Oh, D. H., Kang, J.-K., Kim, S.-Y., Jeon, Y. J., Lee, J. W. and Kim, K. (2003). Facile synthesis of cucurbit[n]uril derivatives via direct functionalization: Expanding utilization of cucurbit[n]uril, *J. Am. Chem. Soc.*, **125**, pp. 10186–10187.

9. Kim, K., Selvapalam, N., Ko, Y. H., Park, K., Kim, D. and Kim. J. (2007). Functionalized cucurbiturils and their applications, *Chem. Soc. Rev.*, **36**, pp. 267–279.

10. Cui, S.-C., Tachikawa, T., Fujitsuka, M. and Majima, T. (2011). Photoinduced electron transfer in a quantum dot-cucurbituril supramolecular complex, *J. Phys. Chem. C*, **115**, pp. 1824–1830.

11. An, Q., Li, G., Tao, C., Li, Y., Wu, Y. and Zhang, W. (2008). A general and efficient method to form self-assembled cucurbit[n]uril monolayers on gold surfaces, *Chem. Commun.*, **17**, pp. 1989–1991.

12. Young, J. F., Nguyen, H. D., Yang, L., Huskens, J., Jonkheijm, P. and Brunsveld, L. (2010). Strong and reversible monovalent supramolecular protein immobilization, *ChemBioChem*, **11**, pp. 180–183.

13. Mahajan, S., Lee, T.-C., Biedermann, F., Hugall, J. T., Baumberg, J. J. and Scherman, O. A. (2010). Raman and SERS spectroscopy of cucurbit[n]urils, *Phys. Chem. Chem. Phys.*, **12**, pp. 10429–10433.

14. Lee, T.-C. and Scherman, O. A. (2010). Formation of dynamic aggregates in water by cucurbit[5]uril capped with gold nanoparticles, *Chem. Commun.*, **46**, pp. 2438–2440.

15. Cao, M., Lin, J., Yanga, H. and Cao, R. (2010). Facile synthesis of palladium nanoparticles with high chemical activity using cucurbit[6] uril as protecting agent, *Chem. Commun.*, **46**, pp. 5088–5090.

16. Premkumar, T., Lee, Y. and Geckeler, K. E. (2010). Macrocycles as a tool: A facile and one-pot synthesis of silver nanoparticles using cucurbituril designed for cancer therapeutics, *Chem. Eur. J.*, **16**, pp. 11563–11566.

17. Premkumar, T. and Geckeler, K. E. (2010). Cucurbit[7]uril as a tool in the green synthesis of gold nanoparticles, *Chem. Asian. J.*, **5**, pp. 2468–2476.

18. de la Rica, R. and Velders, A. H. (2010). Supramolecular Au nanoparticle assemblies as optical probes for enzyme-linked immunoassays, *Small*, **7**, pp. 66–69.

19. de la Rica, R. and Velders, A. H. (2011). Biomimetic crystallization of Ag2S nanoclusters in nanopore assemblies, *J. Am. Chem. Soc.*, **133**, pp. 2875–2877.

20. Barooah, N., Bhasikuttan, A. C., Sudarsan, V., Choudhury, S. D., Pala, H. and Mohanty, J. (2011). Surface functionalized silver nanoparticle conjugates: Demonstration of uptake and release of a phototherapeutic porphyrin dye, *Chem. Commun.*, **47**, pp. 9182–9184.

21. Lu, X. and Masson, E. (2011). Formation and stabilization of silver nanoparticles with cucurbit[n]urils (n = 5–8) and cucurbituril-based pseudorotaxanes in aqueous medium, *Langmuir*, **27**, pp. 3051–3058.

22. Taylor, R. W.; Lee, T.-C., Scherman, O. A., Esteban, R., Aizpurua, J., Huang, F. M., Baumberg, J. J. and Mahajan, S. (2011). Precise subnanometer plasmonic junctions for SERS within gold nanoparticle assemblies using cucurbit[n]uril "glue", *ACS Nano*, **5**, pp. 3878–3887.

23. Tao, C.-A., An, Q., Zhu, W., Yang, H., Li, W., Lin, C., Xu, D. and Li, G. (2011). Cucurbit[n]urils as a SERS hot-spot nanocontainer through bridging gold nanoparticles, *Chem. Commun.*, **47**, pp. 9867–9869.

24. Kim, C., Agasti, S. S., Zhu, Z., Issacs, L. and Rotello, V. M. (2010). Recognition-mediated activation of therapeutic gold nanoparticles inside living cells, *Nat. Chem.*, **2**, pp. 962–966.

25. Coulston, R. J., Jones, S. T., Lee, T.-C., Appel, E. A. and Scherman, O. A. (2011). Supramolecular gold nanoparticle–polymer composites formed in water with cucurbit[8]uril, *Chem. Commun.*, **47**, pp. 164–166.

26. Zhang, J., Coulston, R. J., Jones, S. T., Geng, J., Scherman, O. A. and Abell, C. (2012). One-step fabrication of supramolecular microcapsules from microfluidic droplets, *Science*, **335**, pp. 690–694.

27. Stavens, K. B., Pusztay, S. V., Zou, S., Andres, R. P. and Wei, A. (1999). Encapsulation of neutral gold nanoclusters by resorcinarenes, *Langmuir*, **15**, pp. 8337–8339.

28. Tshikhudo, T. R., Demuru, D., Wang, Z., Brust, M., Secchi, A., Arduini, A. and Pochini, A. (2005). Molecular recognition by calix[4]arene-modified gold nanoparticles in aqueous solution, *Angew. Chem.*, **117**, pp. 2973–2976.

29. Huang, T., Meng, F. and Qi, L. (2009). Facile synthesis and one-dimensional assembly of cyclodextrin-capped gold nanoparticles and their applications in catalysis and surface-enhanced Raman scattering, *J. Phys. Chem. C*, **113**, pp. 13636–13642.

30. Liu, J., Mendoza, S., Roman, E., Lynn, M. J., Xu, R. and Kaifer, A. E. (1999). Cyclodextrin-modified gold nanospheres. host–guest interactions at work to control colloidal properties, *J. Am. Chem. Soc.*, **121**, pp. 4304–4305.

31. Lee, T.-C. and Scherman, O. A. (2012). A facile synthesis of dynamic supramolecular aggregates of cucurbit[n]uril (n=5–8) capped with gold nanoparticles in aqueous media, *Chem.–Eur. J.*, **18**, pp. 1628–1633.

32. Chio, W.-I. K., Davison, G., Jones, T., Liu, J., Parkin, I. P., Lee and T.-C. (2020). Quantitative SERS detection of uric acid via formation of precise plasmonic nanojunctions within aggregates of gold nanoparticles and cucurbit[n]uril, *J. Vis. Exp.*, **164**, e61682.

33. Chio, W.-I. K, Peveler, W. J., Assaf, K. I., Moorthy, S., Nau, W. M., Parkin, I. P., Olivo, M. and Lee, T.-C. (2019). Selective detection of nitroexplosives using molecular recognition within self-assembled plasmonic nanojunctions, *J. Phys. Chem. C*, **123**, pp. 15769–15776.

34. Chio, W.-I. K, Moorthy, S., Perumal, J., Dinish, U. S., Parkin, I. P., Olivo, M. and Lee, T.-C. (2020). Dual-triggered nanoaggregates of cucurbit[7]uril and gold nanoparticles for multi-spectroscopic quantification of creatinine in urinalysis, *J. Mater. Chem. C*, **8**, pp. 7051–7058.

35. Asim, M. H., Ijaz, M., Rösch, A. C. and Bernkop-Schnürch, A. (2020). Thiolated cyclodextrins: News perspectives for old excipients, *Coord. Chem. Rev.*, **420**, 213433.

36. Wei, A. (2005). Calixarene-encapsulated nanoparticles: Self-assembly into functional nanomaterials, *Chem. Commun.*, **2006**, pp. 1581–1591.

37. Biedermann, F., Uzunova, V. D., Scherman, O. A., Nau, W. M. and De Simone, A. (2012). Release of high-energy water as an essential driving force for the high-affinity binding of cucurbit[n]urils, *J. Am. Chem. Soc.*, **134**, pp. 15318–15323.

38. Lambert, H., Mohan, N. and Lee, T.-C. (2019). Ultrahigh binding affinity of a hydrocarbon guest inside cucurbit[7]uril enhanced by strong host–guest charge matching, *Phys. Chem. Chem. Phys*, **21**, pp. 14521–14529.

39. Klajn, R., Stoddart, J. F. and Grzybowski, B. A. (2010). Nanoparticles functionalised with reversible molecular and supramolecular switches, *Chem. Soc. Rev.*, **39**, pp. 2203–2237.

40. Lee, C., Lawrie, B., Pooser, R., Lee, K.-G., Rockstuhl, C. and Tame, M. (2021). Quantum plasmonic sensor, *Chem. Rev.*, **121**, pp. 4743–4804.

41. Li, M., Cushing, S. K. and Wu, N. (2015). Plasmon-enhanced optical sensors: A review, *Analyst*, **140**, pp. 386–406.

42. Brolo, A. G. (2012). Plasmonics for future biosensors, *Nature Photonics*, **6**, pp. 709–713.

43. Nocerino, V., Miranda, B., Tramontano, C., Chianese, G., Dardano, P., Rea, I. and De Stafano, L. (2022). Plasmonic nanosensors: Design, fabrication, and applications in biomedicine, *Chemosensors*, **10**, 150.

44. Miller, M. M. and Lazarides, A. A. (2005). Sensitivity of metal nanoparticle surface plasmon resonance to the dielectric environment, *J. Phys. Chem. B*, **109**, pp. 21556–21565.

45. Jeong, H.-H., Mark, A. G., Alarcón-Correa, M., Kim. I., Osward, P., Lee, T.-C. and Fischer, P. (2016). Dispersion and shape engineered plasmonic nanosensors, *Nat. Commun.*, **7**, 11331.

46. Lee, K.-S. and El-Sayed, M. A. (2006). Gold and silver nanoparticles in sensing and imaging: Sensitivity of plasmon response to size, shape, and metal composition, *J. Phys. Chem. B*, **110**, pp. 19220–19225.

47. Yguerabide, J. and Yguerabide, E. E. (1998). Light-scattering submicroscopic particles as highly fluorescent analogs and their use as tracer labels in clinical and biological applications: II. Experimental characterization, *Anal. Biochem.*, **262**, pp. 157–176.

48. Pitt, J. J. (2009). Principles and applications of liquid chromatography-mass spectrometry in clinical biochemistry, *Clin. Biochem. Rev.*, **30**, pp. 19–34.

49. Valasek, M. A. and Repa, J. J. (2005). The power of real-time PCR, *Adv. Physiol. Educ.*, **29**, pp. 151–159.

50. Alhajj, M. and Farhana, A. (2022). *Enzyme Linked Immunosorbent Assay* (StatPearls Publishing LLC).

51. Koczula, K. M. and Gallotta, A. (2016). Lateral flow assays, *Essays Biochem.*, **60**, pp. 111–120.

52. Hussain, C. M. and Kecili, R. (2020). "Electrochemical techniques for environmental analysis." *Modern Environmental Analysis Techniques for Pollutants*, pp. 199–222 (Elsevier, Netherlands).

53. Yoo, E.-H. and Lee, S.-Y. (2010). Glucose biosensors: An overview of use in clinical practice, *Sensors (Basel)*, **10**, pp. 4558–4576.

Chapter 2

Host–Guest Chemistry of Macrocycles

This chapter is intended to introduce selected macrocyclic hosts (cavitands), including calixarenes, cyclodextrins, cucurbiturils, and pillararenes. We will also discuss the state-of-the-art synthesis and functionalisation of cyclodextrins and cucurbiturils, as well as their fundamental properties and host–guest chemistry.

2.1 Macrocyclic Hosts

Supramolecular chemistry describes the non-covalent interactions between two or more molecules or ions to form a structural complex. Macrocyclic hosts such as calixarenes, cyclodextrins, cucurbiturils, and pillararenes (Fig. 2.1), which can trap small analyte molecules or ions inside their cavities to form host–guest complexes, play a critical role in supramolecular chemistry, attracting great attention for their wide-range applications including but not limited to sensing, drug delivery, molecular machines, and catalysis.

Macrocycle-Functionalised Nanosensors
Weng-I Katherine Chio and Tung-Chun Lee

ISBN 978-981-5129-33-5 (Hardcover), 978-1-003-51057-4 (eBook)
www.jennystanford.com

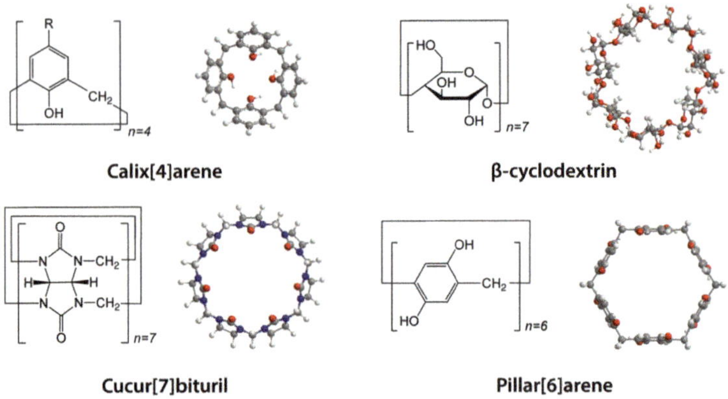

Figure 2.1 Chemical structures and molecular models (top view) of different macrocycles. (Colour code: grey = C atom, white = H atom, red = O atom, and blue = N atom.)

2.1.1 Calixarenes

Calix[*n*]arenes (CXs) are basket-shaped macrocycles formed through oligomerisation of phenols and formaldehyde. Although pioneered by Adolf von Baeyer in 1872, the potential of calixarenes as a host molecule analogous to enzymes was first realised by John Cornforth in 1955. The word calixarene was coined by C. David Gutsche in 1978 for the resemblance to the shape of the calyx krater vases in ancient Greece [1]. CXs are characterised by a wide upper rim and a narrow lower rim. As the native forms of CXs are insoluble in water, they need to be functionalised with groups such as carboxylic acids, sulfonates, phosphonates, or peptides to attain aqueous solubility [2]. CXs are able to form supramolecular assemblies and capsules in both solution and crystalline states, as illustrated by the hexameric "Archimedean solid" nano-capsules held together by 48 intermolecular hydrogen bonds between the phenoxy groups on the lower rim [3, 4]. Their cation-receptor properties mediated by π-donation of the phenyl rings allow them to be utilised in ion-sensitive electrodes and sensors [5]. In addition, they have also been employed in the synthesis of supramolecular polymers and fullerene complexes, as well as in the applications of enzyme mimetics, nonlinear optics, and supramolecular catalysis. Nevertheless, larger aqueous macromolecular architectures involving CXs are relatively rare, probably owing to their low affinity constants.

2.1.2 Cyclodextrins

Cyclodextrins (CDs) are cyclic oligosaccharides consisting of six (α-CD), seven (β-CD), or eight (γ-CD) glucose monomer units. α-CD and β-CD were first described by A. Villiers in 1891 but their chemistry was first identified by F. Schardinger in the early 20th century. Later, the term "cyclodextrins" was suggested by F. Cramer in the 1940s while γ-CD was discovered by K. J. Freudenberg in 1948 [6]. The naturally occurring CDs have been shown to form stable aqueous complexes with many chemicals for a century.

CDs are water-soluble host molecules with a hydrophobic cavity which enables them to accommodate a large variety of organic molecules with an average binding constant of 102.5 ± 1.1 M^{-1} (based on 1257 reported binding affinities) [7]. Host–guest complexes with higher binding constants such as β-CD and adamantane ($K_a = 10^4$ M^{-1}) have successfully been employed in the construction of supramolecular polymers [8]. Meanwhile, CDs have been utilised for drug delivery purposes where hydrophobic drugs, such as praziquantel, can be complexed and solubilised by CDs. The encapsulated drugs can then penetrate body tissues and be released under specific conditions [9].

On the other side, a supramolecular bioconjugate between two model proteins (enhanced cyan fluorescent protein and enhanced yellow fluorescent protein) has also been demonstrated. The two fluorescent proteins were first functionalised with the β-CD host and the lithocholic acid guest, respectively. The formation of hetero-dimers was then mediated by the complexation of the pendant host–guest pairs, resulting in a strong fluorescence resonance energy transfer (FRET) effect [10]. It is noteworthy that CDs (α-, β-, and γ-) are all generally recognised as safe by the U.S. Food and Drug Administration (FDA), allowing their rather widespread applications in food industries (as cholesterol scavengers) and bio-medical fields.

Last but not least, ease in the chemical functionalisation of CDs leads to a rich variety of accessible CD derivatives, including enzyme mimics and molecular machines.

2.1.3 Cucurbiturils

In 1905, contemporaneous with the pioneering work of Schardinger on cyclodextrins, Behrend and co-workers reported that the condensation of glycoluril (acetylene urea) and formaldehyde in concentrated HCl yielded an insoluble polymeric substance known as Behrend's polymer [11]. The molecular structure of this substance remained unclear until 1981 when Mock and co-workers reinvestigated Behrend's report [12]. By using X-ray crystallography, it was found that the substance consists of 6 glycoluril units bound together by 12 methylene bridges, which is then known as cucurbit [6] uril (CB6). The name "cucurbituril" was proposed by Mock for the resemblance of the CB molecular structure to a pumpkin, which belongs to the plant family Cucurbitaceae. From 2000 onwards, a whole family of cucurbit[*n*]urils (CB*n*, n = 5–8, 10, 13–15) have been reported with guest binding properties dependent on their different cavity sizes. Chemical functionalisation of CBs has also become achievable, providing an effective handle for CBs to be appended onto functional surfaces and within polymeric networks.

CBs are barrel-shaped macrocyclic host molecules, which are highly rigid and symmetric [13]. Having a well-defined and unique structure (the presence of a central hydrophobic cavity with two tighter carbonyl portal openings), CBs outweigh the classical supramolecular hosts for their higher selectivity towards the shape and charge of guest molecules [14]. Among all, CB7 stands out for its remarkably high binding affinity for neutral guests in water (similar to the biotin-avidin pair) whereas similar binding affinities can only be obtained for host–guest complexation between molecules of high and opposite charges in other supramolecular systems [15].

CBs have been employed in constructing a wide range of supramolecular architectures, including catenanes, pseudo-rotaxanes, and polyrotaxanes, as well as in the preparation of supramolecular polymers and protein conjugates. Because of their low toxicity, CBs can potentially be used in various biological applications, such as enzyme assays and drug delivery. On the other hand, CBs can potentially act as nano-reactors in aqueous environment. Initial examples of chemoselective and in part catalytic reactions inside CBs in water have already become available. Nonetheless, CBs are particularly useful for sensing for their capability to form a wide variety of host–guest complexes with high binding affinity.

2.1.4 Pillararenes

Pillar[n]arenes (PAs) are a relatively new family of pillar-shaped macrocycles composed of hydroquinone linked by methylene bridges at *para* positions which can be relatively easy to functionalise. They are structurally similar to CXs while being more rigid and symmetrical. The synthesis of PA5 via Lewis acid catalysed condensation of 1,4-dimethoxybenzene and paraformaldehyde was first reported by T. Ogoshi in 2008, which has attracted great interest in the past decade [16]. Interestingly, PA exhibits conformational properties and planar chirality due to the presence of hydroquinone oxygens on both rims, with symmetric PA5 and PA6 being the most commonly used homologues [17]. Native PAs are readily soluble in organic solvents and form a wide range of host–guest complexes by trapping the guest molecules inside their hydrophobic cavities, making them complementary to the more water-soluble cavitands like CDs and CBs. With recent and continuous research efforts, PAs have found potential applications in molecular machines, drug delivery, catalysis, and sensing.

2.2 Preparation of Macrocycles

Over the years, there have been numerous methods developed for the synthesis and separation of cavitands. Despite the ease in synthesis, it is somehow challenging to purify the macrocycles after synthesis due to the co-existence of different members in the product. In particular, we will focus on the synthesis and purification of CDs and CBs in the following subsections.

2.2.1 Synthesis of Cyclodextrins

While CDs are naturally occurring cyclic oligosaccharides with the first being discovered in 1891 (Table 2.1), various industrial methods have been developed to synthesise them on a large scale. Typically, CDs are synthesised by enzymatic conversion of liquefied starch in the presence of cyclodextrin glucanotransferase (CGTases, EC 2.4.1.19) such as alkaliphilic *Bacillus* species (Fig. 2.2) [18]. A mixture of CD homologues is formed with the ratio being dependent on the strain of CGTases and reaction conditions. The yield of CDs is affected by the degradation of CDs into the linear oligosaccharides and product inhibition of enzymes, which can be overcome by adding complexing agents (e.g., toluene, ethyl alcohol, acetone) to precipitate the CDs [18, 19]. Note that the use of complexing agents can also help to mainly precipitate a specific CD homologue. Nevertheless, the CD homologues can be isolated based on their water solubility or using chromatography, with β-CD being easiest to purify due to its low solubility (see Section 2.4.1).

Table 2.1 Timeline of the discovery of homologues in the main CD family (CD derivatives are excluded)

Year	CD homologue	Research group
1891	α-, β-CD (described)	Viller
1905	α-, β-CD (identified)	Schardinger
1948	γ-CD	Frendensberg

2.2.2 Synthesis of Cucurbiturils

The first CB homologue (CB6) was synthesised by Behrend and co-workers as a condensation product of glycoluril and formaldehyde in 1905. A century later, the research groups of Kim and Day reported the synthesis and isolation of other CB homologues, including CB5, CB7, CB8, and CB10·CB5 (i.e., CB5 encapsulated within the cavity of CB10), with a different number of glycoluril units (Fig. 2.3), by modifying the reaction conditions such as reagent (H_2SO_4/HCl), temperature (80–100°C), and time (10–100 h) [20–24]. The isolation of CB homologues is based on their extent of solubilities in water, water/methanol, and diluted HCl, with CB6 being the major product.

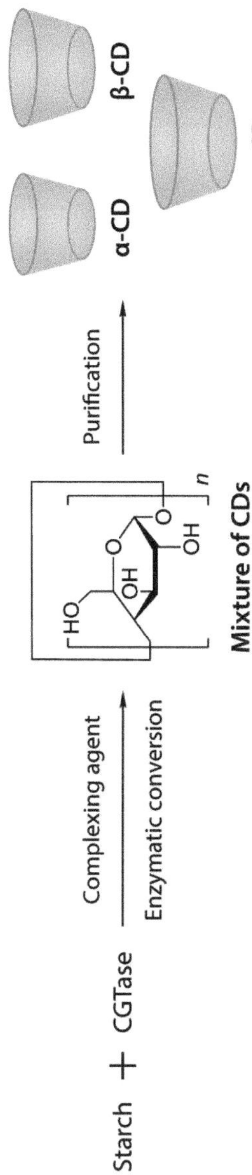

Figure 2.2 General synthesis and isolation of CDs, showing the CD homologues (α, β, and γ-CD).

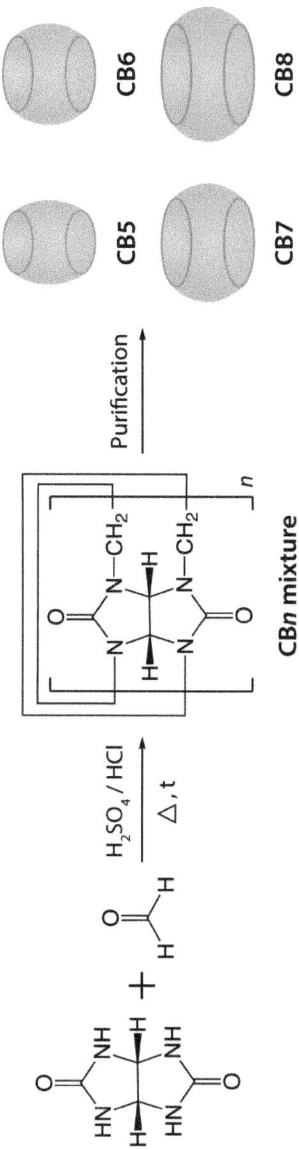

Figure 2.3 General synthesis and isolation of CBs, showing the four smallest CB homologues (CB5-8).

In addition, CB10 can be isolated from CB10·CB5 by competitive guest binding to displace CB5 and subsequent removal of the newly encapsulated guests (Fig. 2.4). Such isolation was first performed by Isaacs and co-workers who used an excess amount of melamine diamine to form a 2:1 complex with CB10 [25, 26]. They removed the two guests by washing with methanol and reacting with acetic anhydride, followed by washing with methanol, dimethyl sulfoxide, and water. A few years later, Day and co-workers reported a simpler approach, in which they used the commercially available 1,12-diaminododecane to displace CB5 at low pH and subsequently removed it by repetitive washing with NaOH/methanol solution, before recrystallising CB10 from HCl [27].

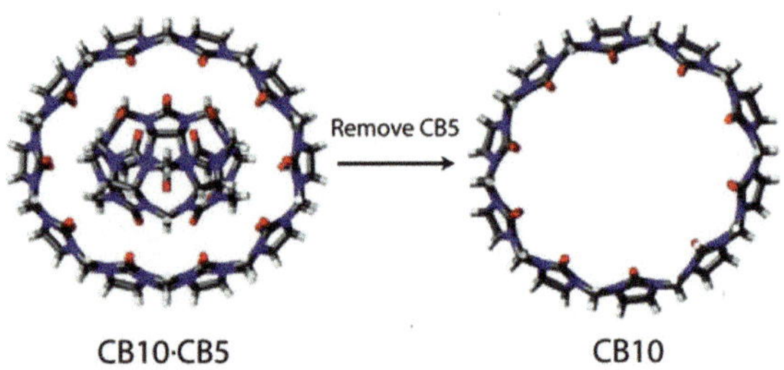

Figure 2.4 X-ray crystal structure of CB10·CB5 and CB10 [26].

The larger members of the CB family, i.e., CB13, 14, 15, which adopt twisted conformations, were isolated in 2013 and 2016 (Fig. 2.5, Table 2.2) [28, 29]. The discovery of the twisted CB13–15 implies that there is a trace amount of larger CB homologues formed during CB synthesis, which is still based on the acid-catalysed condensation of glycoluril and formaldehyde reported by Behrend.

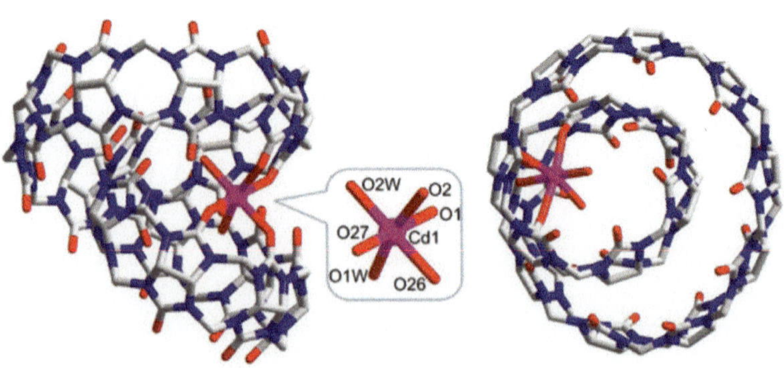

Figure 2.5 Side (left) and top view (right) of the X-ray crystal structure of Cd^{2+}/CB15 complex. CB15 is the current largest member of the CB family [29].

Table 2.2 Timeline of the discovery of homologues in the main CB family (CB derivatives are excluded) [24]

Year	CB homologue	Research group
1905	CB6	Behrend
1981	CB6	Mock
2000	CB5,7,8	Kim
2002	CB10·CB5	Day
2005	CB10	Isaacs
2013	CB14	Tao
2016	CB13,15	Tao

2.3 Functionalisation of Macrocycles

While the major family members of the macrocycles offer some important applications, the versatilities of the macrocycles can be further enhanced through functionalisation. For instance, the properties of CDs and CBs (e.g., solubility and reactivity) can be tailored by attaching different functional groups to them, to allow more efficient host–guest complexation for nanosensing.

2.3.1 Functionalisation of Cyclodextrins

While some functionalised CDs are commercially available, there are various approaches used to functionalise CDs in research, such as the capping strategies and site-oriented selective functionalisation for the primary rim and supramolecular inclusion complex strategies and bridging strategies for the secondary rim [30]. CDs are functionalised by attaching different functional groups to the hydroxyl groups of the glucose units at 2-, 3-, and 6-positions (Fig. 2.6). However, the selective modification of the CD structures is somehow challenging due to the competition between the substitution sites [30].

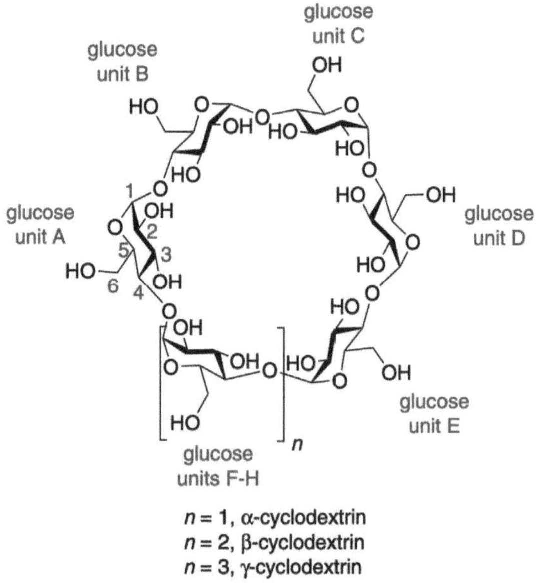

Figure 2.6 Structure of CD homologues showing the position of hydroxyl groups on the glucose units [30].

Fortunately, the hydroxyl groups at the 6-position on the primary (narrower) rim are most reactive for nucleophilic substitutions as they are primary alcohols. For the secondary (wider) rim, the hydroxyl groups at 2-positions are the most acidic (pKa = 12.2) whereas the ones at 3-positions are the least reactive [30]. Thus, the hydroxyl groups of CDs can be persubstituted, randomly substituted, persubstituted at selected positions, selectively substituted, and monosubstituted (Fig. 2.7) [31]. Note that it is much easier to modify a CD that has been substituted with one or more functional groups.

To prepare persubstituted CDs where all hydroxyl groups are replaced by the same functional group, we can react the native CDs with an excess amount of the reagent in the presence of a base. If we reduce the amount of reagent to a certain extent (fewer equivalent than the hydroxyl groups), we end up with randomly substituted CDs, where the hydroxyl groups are replaced at different positions, with the exact ratio being unknown. We can also prepare persubstituted CDs at selected positions, i.e., 2-position, 3-position, 6-position, or any combination of the three substitution sites, by taking the aforementioned different reactivities into account and using a protection or deprotection approach. Moreover, the hydroxyl groups at 2-position, 3-position, or 6-position can be replaced using either direct (i.e., react native CDs with reagents) or indirect synthesis (i.e., deprotect persubstituted CDs) for monosubstituted CDs. Selectively substituted CDs are similar to their monosubstituted counterparts but with two or more substituents. Thus, it is more challenging to functionalise due to a large number of possible combinations.

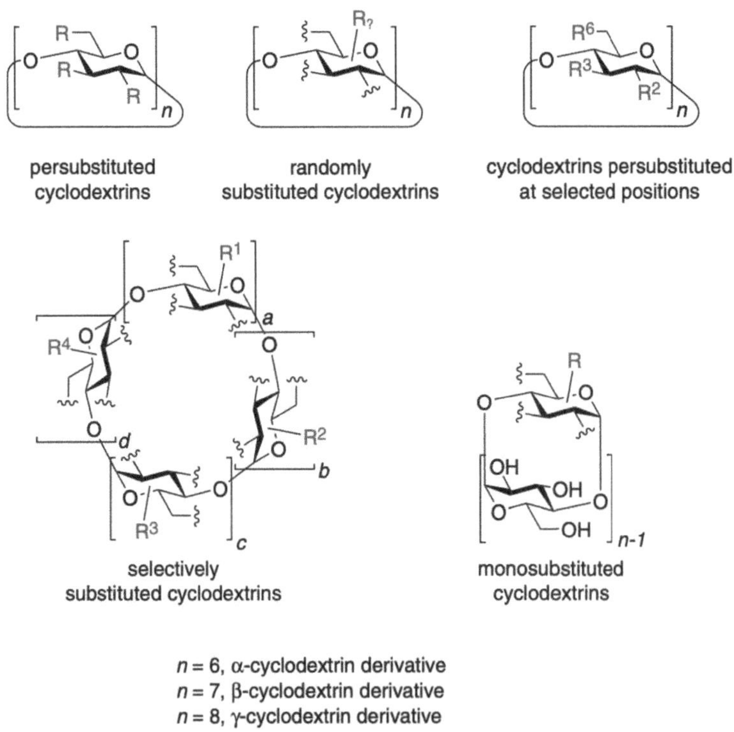

Figure 2.7 Schematic illustration of different CD derivatives [31].

2.3.2 Functionalisation of Cucurbiturils

Unlike other macrocycles such as CDs and CXs, it is quite difficult to attach functional groups on to the molecular framework of CBs. Since the first report of CB*n* derivative, decamethyl-CB5 (full equatorially methylated CB5), by Stoddart in 1992, various CB*n* derivatives including functionalised CB*n*, inverted CB*n* (*i*CB*n*), nor-*seco*-CB*n*, and congeners (e.g., bambus[6]uril and hemicucurbit[*n*]uril) have also been produced.

Functionalised CBs can be produced by either synthesis through glycoluril derivatives or direct functionalisation, with the H atoms at the equatorial position being fully substituted, partially substituted, or monosubstituted (Fig. 2.8) [32]. For instance, we can react glycoluril derivatives carrying substituents R at the equatorial positions with formaldehyde to form a fully substituted CB in a condensation reaction. It is also possible to form a mixture of partially substituted CBs and native CBs by reacting a mixture of glycoluril and substituted glycoluril with formaldehyde. In addition, fully hydroxylated CBs at the equatorial positions can be directly functionalised by oxidation of native CBs under harsh conditions (hot $K_2S_2O_8$) while partially hydroxylated CBs can also be formed with a complex mixture of hydroxylated products [32]. Note that the hydroxylated CBs allow relatively easier further modification with different functional groups, usually with an enhanced solubility in organic solvents such as DMSO and DMF [24, 32].

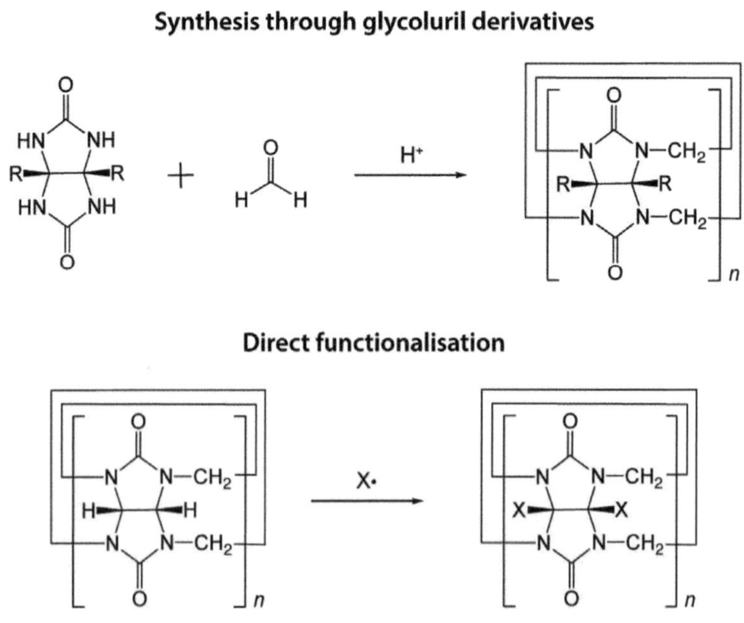

Figure 2.8 Schematic illustration showing different routes to functionalise CBs.

2.4 Properties of Macrocycles

The properties of macrocycles are governed by their characteristic size and shape, arrangement of atoms in the molecular framework, and orientation of functional groups. A fundamental understanding of the physical and chemical properties is necessary to tailor the macrocycles for particular applications.

2.4.1 Properties of Cyclodextrins

CDs are truncated cone-shaped molecules with two hydroxyl rims of different sizes (Fig. 2.9). The narrow rim contains *n* primary hydroxyl groups while the wider rim contains 2*n* secondary hydroxyl groups. The electrostatic potential map shows the cavity of CDs is relatively more hydrophobic (nonpolar) than the hydrophilic (polar) outer surface due to the presence of hydroxyl groups at the surface (Fig. 2.10) [33, 34]. This allows CDs to form non-covalent inclusion complexes inside their hydrophobic cavity with a preference for neutral or anionic guests. The heights of the CD homologues are the same (0.79 nm) whereas other parameters, including portal diameters, cavity diameters, outer diameters, cavity diameters, and thus volume, increase with the number of glucopyranose units (Table 2.3). Note that CDs with an even number of glucopyranose units are highly soluble in water at room temperature while the one with an odd number is less water soluble.

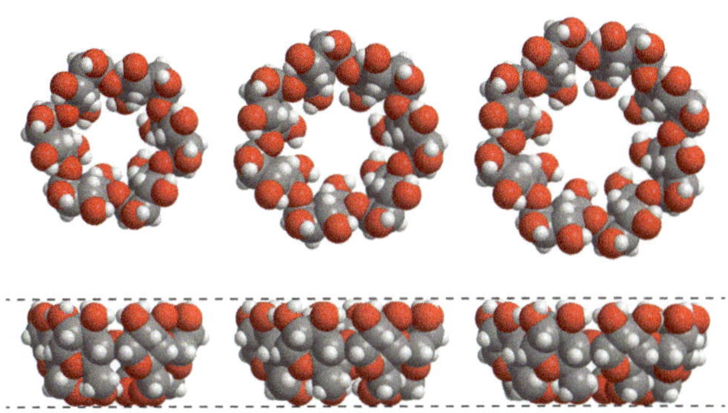

Figure 2.9 Molecular structures (top and side view) of native CDs, with increasing cavity size in order of α-, β-, and γ-CD.

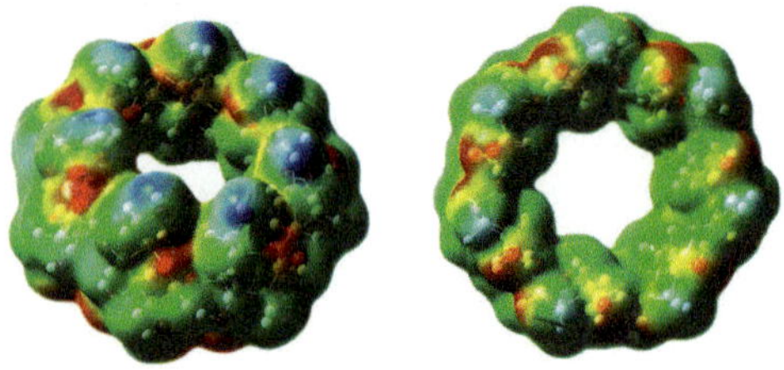

Figure 2.10 Electrostatic potential map (side and top view) of *β*-CDs. The red, green, and blue colours indicate the negative, neutral, and positive regions respectively [34].

Table 2.3 Structural parameters and properties of CDs [35]

	α-CD	*β*-CD	*γ*-CD
No. of glucopyranose units	6	7	8
Formula	$C_{36}H_{60}O_{30}$	$C_{42}H_{70}O_{35}$	$C_{48}H_{80}O_{40}$
Molecular weight (g/mol)	972.9	1135.0	1297.1
Cavity diameter (Å)	4.7–5.3	6.0–6.5	7.5–8.3
Outer diameter (Å)	14.6	15.4	17.5
Height (Å)	7.9	7.9	7.9
Cavity volume (Å^3)	174	262	427
Water solubility at 25°C (mM)	145	18.5	232

The probability of CDs forming inclusion complexes can be estimated from the packing coefficient, which is the volume ratio of the guest molecule to the cavity of the macrocycle [36]. Interestingly, an optimal packing coefficient of 55% gives the best binding affinity for host–guest complexes in solution, with a lower or higher value leading to lower binding affinity [36]. Depending on the cavity volume, *n*-CD can host between 6 and 14 water molecules using 55% packing coefficients, which is very similar to the results obtained from molecular dynamics simulations (Table 2.4). The driving forces for formation of CD host–guest complexes include release of high-energy water, solvent effects, hydrophobic interactions, hydrogen bonding, van der Waals interactions, and dipole–dipole interactions [35].

Table 2.4 Occupancy of the cavity of CDs by water molecules. The number of water molecules in the cavity was calculated by assuming an ideal 55% packing coefficient and each water molecule has a volume of 17 Å^3 [37, 38]

	α-CD	β-CD	γ-CD
55% packing coefficient	6	8	14
Molecular dynamics simulation	5	9	14

2.4.2 Properties of Cucurbiturils

CBs are rigid, highly symmetric barrel-shaped molecules, with two negatively charged carbonyl portals. The functionality of CBs as cation-receptors is due to the presence of electron-rich carbonyl oxygens at both portals as illustrated in the electrostatic potential map (Fig. 2.11) [24]. The non-dipolar nature of CBs and the lack of functional groups and lone electron pairs pointing towards the inside of the cavity lead to limited hydrogen bonding interactions and thus a highly hydrophobic cavity [24, 38]. As the electrostatic potentials at the portals and inside the cavity of CBs are significantly more electronegative than CDs, CBs have a distinct preference to bind to cationic guests over neutral guests and even more over anionic guests [13].

Figure 2.11 Electrostatic potential map (top and side view) of CB7, showing the negatively charged carbonyl regions (red) and the electron-deficient equatorial region (blue) [24].

Similar to other macrocycles, major CB homologues have the same height (0.91 nm) whereas other structural parameters, including cavity diameter, outer diameter, and cavity volume, increase systematically with the number of glycoluril units (Fig. 2.12, Table 2.5). The CB portals are approximately 0.2 nm smaller than the equatorial regions, restricting the size of guests binding to CB. Interestingly, CB*n* with an odd number of *n* (CB5 and CB7) have fairly good water solubility of 20–30 mM (similar to *β*-CD) but those with an even number of *n* (CB6 and CB8) have poor water solubility. The solubility of CBs can be improved in acidic solutions, in the presence of metal cations, or upon the formation of inclusion complexes with amphiphilic guests [23, 24].

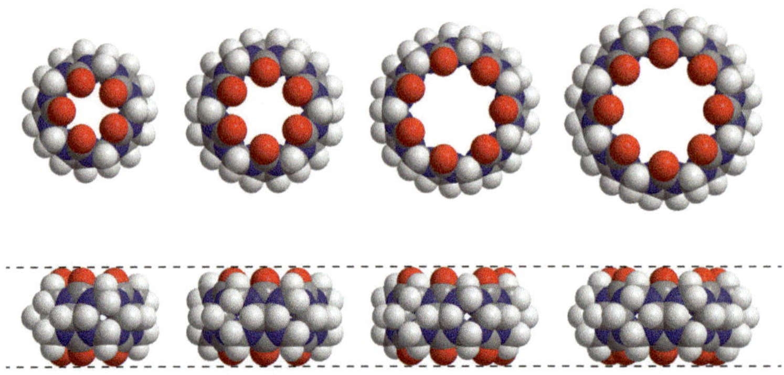

Figure 2.12 Molecular structures (top and side view) of major CB*n* (*n* = 5–8) homologues, with increasing cavity size from CB5 to CB8.

Table 2.5 Structural parameters and properties of CBs [24, 38]

	CB5	CB6	CB7	CB8
No. of glycoluril units	5	6	7	8
Formula	$C_{30}H_{30}N_{20}O_{10}$	$C_{36}H_{36}N_{24}O_{12}$	$C_{42}H_{42}N_{28}O_{14}$	$C_{48}H_{48}N_{32}O_{16}$
Molecular weight (g/mol)	830.7	996.8	1163.0	1329.1
Cavity diameter (Å)	2.4–4.4	3.9–5.8	5.4–7.3	6.9–8.8
Outer diameter (Å)	13.1	14.1	16.0	17.5
Height (Å)	9.1	9.1	9.1	9.1
Cavity volume (Å^3)	68	142	242	367
Water solubility at 25°C (mM)	20–30	0.018	20–30	<0.01

As in the case of CDs, the likelihood of CB to form a host-guest inclusion complex can be also estimated from the packing coefficient. Depending on the inner cavity volume, CB*n* can host between 2 and

12 water molecules as estimated from the 55% PC, and similar results were obtained from molecular dynamics simulations (Table 2.6).

Table 2.6 Occupancy of the inner cavity of CBs by water molecules. The number of water molecules in the cavity was calculated by assuming an ideal 55% packing coefficient and each water molecule has a volume of 17 Å^3 [38]. *The potential energy difference for removal of all water molecules in the CB cavity and transfer of those to a spherical cavity in bulk water [24, 38]

	CB5	CB6	CB7	CB8
55% packing coefficient	2	4	8	12
Molecular dynamic simulations	2	4	7	10
ΔE_{pot}^* (kJ/mol)	−41.6 ± 28.8	−51.1 ± 29.0	−102.4 ± 31.3	−66.2 ± 10.7

Due to the weak dispersion interactions and deficiency of hydrogen bonding, the water molecules inside the CB cavity are of high energy (i.e., enthalpically and entropically unfavourable) [24]. The release of these high-energy water molecules is thus the major driving force for the formation of CB inclusion complexes in aqueous solution (Fig. 2.13). The gain in potential energy for removing all water molecules from the CB cavity is determined by the compromise between the absolute number of water molecules and their engagement in hydrogen bonding [15]. Note that the potential energy of individual water molecules generally decreases with increasing cavity size as there is a higher chance to form hydrogen bonds (Table 2.6). Interestingly, the maximum potential energy is reached for CB7 which does not allow the water molecules within its cavity to arrange in an energetically stable hydrogen-bond network as in the case of CB8 (i.e., structurally similar to bulk water) [15, 24]. This may provide a rationale for the ultra-high binding affinities of CB7 despite having a smaller cavity size than CB8.

2.5 Host–Guest Complexes

The most characteristic property of macrocycles, i.e., the ability to form host–guest complexes with a great variety of guests, makes them particularly suitable for practical applications including, but not limited to, drug delivery, catalysis, and sensing. In this section, we will introduce some famous examples of CD and CB inclusion complexes to show the wide variety of guest molecules available to the macrocycles.

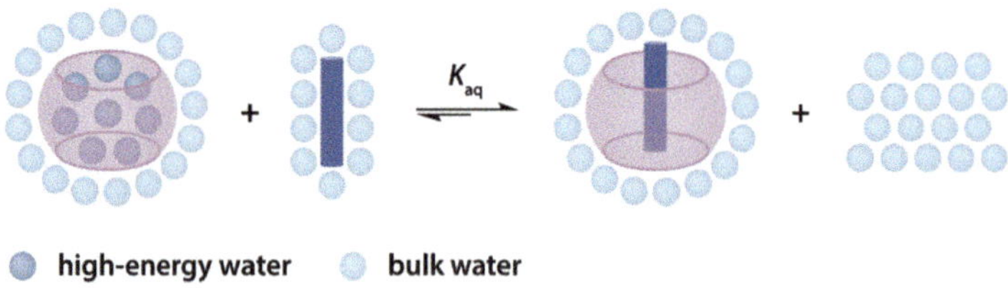

Figure 2.13 The release of high-energy water molecules from the CB cavity upon formation of host–guest complexes.

2.5.1 Host–Guest Chemistry of Cyclodextrins

The very first example of a CD host–guest complex between CD and methylene blue was reported by Cramer in the 1950s, when he observed shifts in the absorption maxima of methylene blue in the presence of CDs. The association-dissociation equilibrium between the unoccupied CD and free guest and the CD inclusion complexes leads to the formation of host–guest complexes. The binding constant is higher when there is size and shape complementarity between the CD cavity and the guest molecule, indicating a more stable inclusion complex with less dissociation.

CDs can form 1:1, 2:1, 1:2, and 2:2 inclusion complexes with the guest molecules depending on their geometries (Fig. 2.14). In general, the smaller native CDs (i.e., α-CD and β-CD) form a 1:1 binary complex with the guest molecules while the larger γ-CD can also form 1:2 inclusion complexes. For example, both α-CD and β-CD can form 1:1 host–guest complexes with phenol and benzoic acid but with different binding geometries [6]. Moreover, β-CD forms a 1:1 inclusion complex with crystal violet and methyl orange whereas γ-CD forms a 1:2 host–guest complex with them [39]. Note that the solubility and stability of guest molecules could be increased upon formation of CD host–guest complexes.

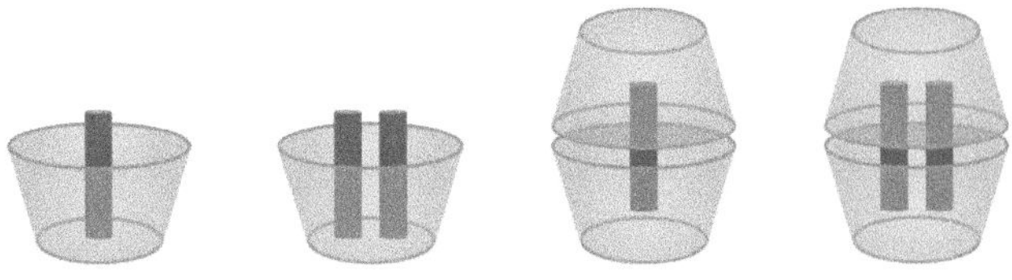

Figure 2.14 Schematic illustrations showing 1:1, 2:1, 1:2, and 2:2 CD host–guest inclusion complexes.

Apart from phenols, α-CD can also bind to aromatic compounds (e.g., nitrobenzene), drugs (e.g., ketoprofen and diclofenac), and diphenyl azo dyes with 1:1 stoichiometry [40–42]. The most commonly used native CD, β-CD, can form 1:1 host–guest complexes with vitamins, drugs (e.g., adamantane and ibuprofen), and flavonoids [43–46]. Furthermore, the larger γ-CD can also form a 1:1 inclusion complex with polycyclic aromatic hydrocarbon (PAH, e.g., naphthalene, anthracene, and corannulene) and antioxidants, while forming both 1:1 and 1:2 host–guest complexes with dyes (1:1 with Congo red but 1:2 with methylene blue) [38, 47–49].

2.5.2 Host–Guest Chemistry of Cucurbiturils

The first CB host–guest complexes between CB6 and alkylammonium and alkyldiammonium ions in aqueous formic acid were reported by Mock and Shih in 1983, just two years after they had identified the molecular structure of CB6 [50]. Since then, the host–guest chemistry of CBs has been widely studied. Note that CBs can also form 1:1, 2:1, 1:2, and 2:2 inclusion complexes with the guest molecules like CDs but also 1:1:1 heteroternary complexes depending on their geometries and charges (Fig. 2.14).

The smaller CB*n* ($n = 5–7$) homologues form a 1:1 binary complex with the guest molecules while the larger CB8 can also form 1:2 and 1:1:1 ternary host–guest complexes (Fig. 2.15). CB5, the smallest member in the CB family, can bind to alkali, alkaline earth, transition metal, and ammonium cations (e.g., Na^+, Ca^{2+}, Fe^{3+}, NH_4^+) and gas molecules (e.g., N_2, O_2, N_2O, CO_2, and CH_4) with 1:1 stoichiometry [51–53]. CB6, the first and most abundant CB member, binds strongly to aliphatic amines (e.g., formation of rotaxanes and pseudorotaxanes) and five-membered aromatic rings (e.g., thiophenes) though CB6 is usually not able to accommodate six-membered aromatic rings [50, 54, 55].

Figure 2.15 Schematic illustrations showing 1:1, 1:2, 1:1:1, 2:1, and 2:2 CB host–guest inclusion complexes.

The relatively larger CB members, CB7 and CB8, have been extensively studied for their unique host–guest binding properties. CB7 has ultra-high binding affinities towards a wide variety of guest molecules including ferrocene, adamantanes, and diamantanes derivatives which have reached and even surpassed that of the biotin-avidin pairs ($\sim10^{15}$ M^{-1}) [56–59]. Thanks to the optimal size and shape complementarity between the rigid CB7 cavity and the diamantine core, the inclusion complex of CB7 and diamantine diammonium ion has remarkable binding constants of 7.2×10^{17} M^{-1} in D_2O and 1.9×10^{15} M^{-1} in 50 mM NaO_2CCD_3 (Fig. 2.16) [58]. Note that CB7 can also bind to a wide range of aromatic guests.

With the cavity volume being ~1.5 times larger than CB7, CB8 is characterised by its ability to host two guest molecules. A stable CB8 heteroternary inclusion complex can be achieved when one guest is electron-rich (π-acceptor) and the other is electron-deficient (π-donor) [60]. For instance, the electron-deficient guest, methyl viologen (MV^{2+}), first binds to CB8 before the electron-rich second guest, 2,6-dihydroxynaphthalene, which cannot bind to CB8 alone, is encapsulated via charge-transfer interactions [61].

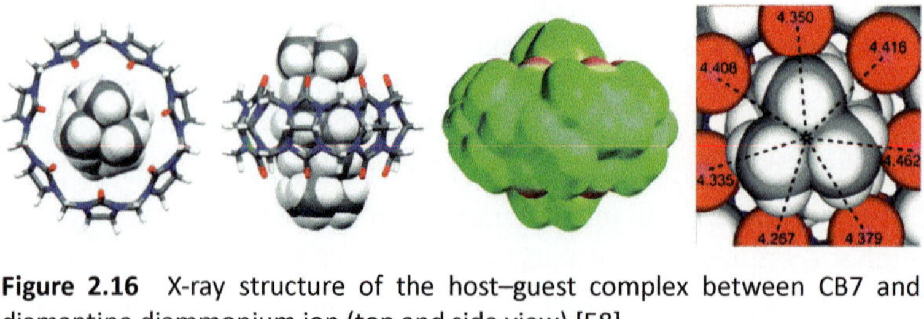

Figure 2.16 X-ray structure of the host–guest complex between CB7 and diamantine diammonium ion (top and side view) [58].

References

1. Gutsche, C. D. and Muthukrishnan, R. (1978). Calixarenes. 1. Analysis of the product mixtures produced by the base-catalyzed condensation of formaldehyde with para-substituted phenols, *J. Org. Chem.*, **43**, pp. 4905–4906.
2. Gutsche, C. D. (2008). *Calixarenes: An Introduction*, 2nd ed. (RSC Publishing).
3. Hof, F., Craig, S. L., Nuckolls, C. and Rebek, J. (2002). Molecular encapsulation, *Angew. Chem. Int. Ed.*, **41**, pp. 1488–1508.
4. Atwood, J. L., Barbour, L. J. and Jerga, A. (2002). Organisation of the interior of molecular capsules by hydrogen bonding, *Proc. Natl. Acad. Sci. U.S.A.*, **99**, pp. 4837–4841.
5. Bohmer, V. (1995). Calixarenes, macrocycles with (almost) unlimited possibilites, *Angew. Chem. Int. Ed.*, **34**, pp. 713–745.
6. Crini, G. (2014). Review: a history of cyclodextrins, *Chem. Rev.*, **114**, pp. 10940–10975.
7. Houk, K. N., Leach, A. G., Kim, S. P. and Zhang, X. (2003). Binding affinities of host–guest, protein-ligand, and protein-transition-state complexes, *Angew. Chem. Int. Ed.*, **42**, pp. 4872–4897.
8. Miyauchi, M. and Harada, A. (2004). Construction of supramolecular polymers with alternating α-, β-cyclodextrin units using conformational change induced by competitive guests, *J. Am. Chem. Soc.*, **126**, pp. 11418–11419.
9. Becket, G., Schep, L. J. and Tan, M. Y. (1999). Improvement of the in vitro dissolution of praziquantel by complexation with alpha-, beta- and gamma-cyclodextrins, *Int. J. Pharm.*, **179**, pp. 65–71.
10. Zhang, L., Wu, Y. and Brunsveld, L. (2007). A synthetic supramolecular construct modulating protein assembly in cells, *Angew. Chem. Int. Ed.*, **46**, pp. 1798–1802.
11. Behrend, R., Meyer, E. and Rusche, F. I. (1905). Ueber condensationsproducte aus glycoluril und formaldehyd, *Justus Liebigs Ann. Chem.*, **339**, pp. 1–37 (in German).
12. Freeman, W. A., Mock, W. L. and Shih, N. Y. (1981). Cucurbituril, *J. Am. Chem. Soc.*, **103**, pp. 7367–7368.
13. Lagona, J., Mukhopadhyay, P., Chakrabarti, S. and Isaacs, L. (2005). The cucurbit[n]uril family, *Angew. Chem. Int. Ed.*, **44**, pp. 4844–4870.
14. Lee, T.-C. and Scherman, O. A. (2010). Formation of dynamic aggregates in water by cucurbit[5]uril capped with gold nanoparticles, *Chem. Commun.*, **46**, pp. 2438–2440.

15. Biedermann, F., Uzunova, V. D., Scherman, O. A., Nau, W. M. and De Simone, A. (2012). Release of high-energy water as an essential driving force for the high-affinity binding of cucurbit[n]urils, *J. Am. Chem. Soc.*, **134**, pp. 15318–15323.

16. Ogoshi, T., Kanai, S., Fujinami, S., Yamagishi, T and Nakamoto, Y. (2008). *para*-Bridged symmetrical pillar[5]arenes: Their lewis acid catalysed synthesis and host–guest property, *J. Am. Chem. Soc.*, **130**, pp. 5022–5023.

17. Shixin, F., Kakuta, T., Yamagishi, T. and Ogoshi, T. (2019). Confomation and planar chirality of pillar[n]arene, *Chem. Lett.*, 48, pp. 1278–1287.

18. Biwer, A., Antranikian, G. and Heinzle, E. (2002). Enzymatic production of cyclodextrins, *App. Microbiol. Biotechnol.*, **59**, pp. 609–617.

19. Leemhuis, H., Kelly, R. M. and Dijukhuizen, L. (2010). Engineering of cyclodextrins glucanotransferases and the impact for biotechnological applications, *App. Microbiol. Biotechnol.*, **85**, pp. 823–835.

20. Kim, J., Jung, I.-S., Kim, S.-Y., Lee, E., Kang, J.-K., Sakamoto, S., Yamaguchi, K. and Kim, K. (2000). New cucurbituril homologues: syntheses, isolation, characterization, and X-ray crystal structures of cucurbit[*n*]uril (*n* = 5, 7, and 8), *J. Am. Chem. Soc.*, **122**, pp. 540–541.

21. Day, A., Arnold, A. P., Blanch, R. J. and Snushall, B. (2001). Controlling factors in the synthesis of cucurbituril and its homologues., *J. Org. Chem.*, **66**, pp. 8094–8100.

22. Day, A. I., Blanch, R. J., Arnold, A. P., Lorenzo, S., Lewis, G. R. and Dance, I. (2002). A cucurbituril-based gyroscane: a new supramolecular form, *Angew. Chem. Int. Ed.*,. **41**, pp. 275–277.

23. Masson, E., Ling, X., Joseph, R., Kyeremeh-Mensah, L. and Lu, X. (2012). Cucurbituril chemistry: a tale of supramolecular success, *RSC Adv.*, **2**, pp. 1213–1247.

24. Assaf, K. I. and Nau, W. M. (2015). Cucurbiturils: from synthesis to high-affinity binding and catalysis. *Chem. Soc. Rev.*, **44**, pp. 394–418.

25. Liu, S., Zavalij, P. Y. and Isaacs, L. (2005). Cucurbit[10]uril, *J. Am. Chem. Soc.*, **127**, pp. 16798–16799.

26. Isaacs, L. (2009). Cucurbit[*n*]urils: from mechanism to structure and function, *Chem. Commun.*, **2009**, pp. 619–629.

27. Pisani, M. J., Zhao, Y., Wallace, L., Woodward, C. E., Keene, F. R., Day, A. I. and Collins, J. G. (2010). Cucurbit[10]uril binding of dinuclear platinum(II) and ruthenium(II) complexes: association/dissociation rates from seconds to hours, *Dalton Trans*, **39**, pp. 2078–2086.

28. Cheng, X.-J., Liang, L.-L., Chen, K., Ji, N.-N., Xiao, X., Zhang, J.-X., Zhang, Y.-Q., Xue, S.-F., Zhu, Q.-J., Ni, X.-L., et al. (2013). Twisted cucurbit[14]uril, *Angew. Chem. Int. Ed.*, **52**, pp. 7252–7255.

29. Li, Q., Qiu, S.-C., Zhang, J., Chen, K., Huang, Y., Xiao, X., Zhang, Y., Li, F., Zhang, Y.-Q., Xue, S.-F., et al. (2016). Twisted cucurbit[*n*]urils, *Org. Lett.*, **18**, pp. 4020–4023.

30. Liu, J., Yu, P., Sollogoub, M. and Zhang, Y. (2019). *Functionalized Cyclodextrins and Their Applications in Biodelivery* (Springer Nature Singapore Pte Ltd.).

31. Rezanka, M. (2018). *Cyclodextrin Fundamentals, Reactivity and Analysis*, eds. Fourmentin, S., Crini, G. and Lichtfouse, E., Chapter 2 "Synthesis of Cyclodextrin Derivatives" (Springer International Publishing), pp. 57–104.

32. Kim, K., Selvapalam, N., Ko, Y. H., Park, K. M., Kim, D. and Kim, J. (2007). Functionalized cucurbiturils and their applications, *Chem. Sov. Rev.*, **36**, pp. 267–279.

33. Connors, K. A. (1997). The stability of cyclodextrin complexes in solution, *Chem. Rev.*, **97**, pp. 1325–1357.

34. Champagne, P. L., Kumar, R., Ling, C.-C. (2018). *Cyclodextrin Applications in Medicine, Food, Environment and Liquid Crystals*, eds. Fourmentin, S., Crini, G. and Lichtfouse, E., Chapter 7 "Supramolecular Liquid Crystals Based on Cyclodextrins" (Springer International Publishing), pp. 183–240.

35. Crini, G., Fourmentin, S., Fenyvesi, E., Torri, G., Fourmentin, M. and Morin-Crini, N. (2018). *Cyclodextrin Fundamentals, Reactivity and Analysis*, eds. Fourmentin, S., Crini, G. and Lichtfouse, E., Chapter 1 "Fundamentals and Applications of Cyclodextrins" (Springer International Publishing), pp. 1–56.

36. Mecozzi, S. and Rebek, J. (1998). The 55 % solution: a formula for molecular recognition in the liquid state, *Chem. Eur. J.*, **4**, pp. 1016–1022.

37. Sandilya, A. A., Natarajan, U. and Priya, M. H. (2020). Molecular view into the cyclodextrin cavity: structure and hydration, *ACS Omega*, **5**, pp. 25655–25667.

38. Nau, W. M., Florea, M. and Assaf, K. I. (2011). Deep inside cucurbiturils: physical properties and volumes of their inner cavity determine the hydrophobic driving force for host–guest complexation, *Isr. J. Chem.*, **51**, pp. 559–577.

39. Hirai, H., Toshima, N. and Uenoyama, S. (1985). Inclusion complex formation of γ-cyclodextrin. One host-two guest complexation with water-soluble dyes in ground state, *Bull. Chem. Soc. Jpn.*, **58**, pp. 1156–1164.
40. Poulson, B. G., Alsulami, Q. A., Sharfalddin, A., Agammy, E. F. E., Mouffouk, F., Emwas, A.-H., Jaremko, L. and Jaremko, M. (2022). Cyclodextrins: Structural, chemical, and physical properties, and applications, *Polysaccharides*, **3**, pp. 1–31.
41. Arancibia, J. A., Boldrini, M. A., Escandar, G. M. (2000). Spectrofluorimetric determination of diclofenac in the presence of α-cyclodextrin, *Talanta*, **52**, pp. 261–268.
42. Abou-Hamdan, A., Bugnon, P., Saudan, C., Lye, P. G. and Merbach, A. E. (2000). High-pressure studies as a novel approach in determining inclusion mechanisms: therodynamnics and kinetics of the host–guest interactions for α-cyclodextrin complexes, *J. Am. Chem. Soc.*, **122**, pp. 592–602.
43. Saha, S., Roy, A., Roy, K. and Roy, M. N. (2016). Study to explore the mechanism to form inclusion complexes of β-cyclodextrin with vitamin molecules, *Sci. Rep.*, **6**, 35764.
44. Paolino, M., Ennen, F., Lamponi, S., Cernescu, M., Voit, B., Cappelli, A., Appelhans, D. and Komber, H. (2013). Cyclodextrin-adamantane host–guest interactions on the surface of biocompatible adamantyl-modified glycodendrimers, *Macromolecules*, **46**, pp. 3215–3227.
45. Cagno, M., Stein, P. C., Skalko-Basnet, N., Brandl, M., Bauer-Brandl, A. (2011). Solubilization of ibuprofen with β-cyclodextrin derivatives: energetic and structural studies, *J. Pharm. Biomed. Anal.*, **55**, pp. 446–451.
46. Rezende, B. A., Cortes, S. F., De Sousa, F. B., Lula, I. S., Schmitt, M., Sinisterra, R. D., Lemos, V. S. (2009). Complexation with β-cyclodextrin confers oral activiity on the flavonoid dioclein, *Int. J. Pharm.*, **367**, pp. 133–139.
47. Ravelet, C., Ravel, A., Grosset, C., Villet, A., Geze, A., Wouessidijewe, D. and Peyrin, E. (2002). Stoichiometry and formation constants of six PAHs with γ-cyclodextrin, determined by HPLC using a cyano station phase, *J. Liq. Chromatogr.*, **25**, pp. 421–432.
48. Joshi, H., Sreejith, S., Dey, R. and Stuparu, M. C. (2016). Host-guet interaction between corannulene and γ-cyclodextrin: mass spectrometric evidence of a 1:1 inclusion complex formation, *RSC Adv.*, **6**, pp. 110001–110003.

49. Anselmi, C., Centini, M., Ricci, M., Buonocore, A., Granata, P., Tsuno, T. and Facino, R. M. (2006). Analytical characterization of a ferulic acid/γ-cyclodextrin inclusion complex, *J. Pharm. Biomed. Anal.*, **40**, pp. 875–881.

50. Mock, W. L. and Shih, N. Y. (1983). Host–guest binding capacity of cucurbituril, *J. Org. Chem.*, **48**, pp. 3618–3619.

51. Buschmann, H.-J., Cleve, E., Jansen, K., Wego, A. and Schollmeyer, E. (2001). Complex formation between cucurbit[*n*]urils and alkali, alkaline earth and ammonium ions in aqueous solution, *J. Incl. Phenom. Macrocycl. Chem.*, **40**, pp. 117–120.

52. Buschmann, H.-J., Cleve, E., Jansen, K. and Schollmeyer, E. (2001). Determination of complex stabilities with nearly insoluble host molecules: cucurbit[5]uril, decamethylcucurbit[5]uril and cucurbit[6]uril as ligands for the complexation of some multicharged cations in aqueous solution, *Anal. Chim. Acta*, **437**, pp. 157–163.

53. Kellersberger, K. A., Anderson, J. D., Ward, S. M., Krakowiak, K. E. and Dearden, D. V. (2001). Encapsulation of N_2, O_2, methanol, or acetonitrile by decamethylcucurbit[5]uril$(NH_4^+)_2$ complexes in the gas phase: influence of the guest on "lid" tightness, *J. Am. Chem. Soc.*, **123**, pp. 11316–11317.

54. Tuncel, D., Cindir, N. and Koldemir, U. (2006). [5]Rotaxane and [5]pseudorotaxane based on cucurbit[6]uril and anchored to a meso-tetraphenyl porphyrin, *J. Incl. Phenom. Macrocycl. Chem.*, **55**, pp. 373–380.

55. Barrow, S. J., Kasera, S., Rowland, M. J., del Barrio, J. and Scherman, O. A. (2015). Cucurbituril-based molecular recognition, *Chem. Rev.*, **115**, pp. 12320–12406.

56. Lee, J. W., Samal, S., Selvapalam, N., Kim, H.-J. and Kim, K. (2003). Cucurbituril homologues and derivatives: new opportunities in supramolecular chemistry, *Acc. Chem. Res.*, **36**, pp. 621–630.

57. Liu, S., Ruspic, C., Mukhopadhyay, P., Chakrabarti, S., Zavalij, P. Y. and Isaacs, L. (2005). The cucurbit[*n*]uril family: prime components for self-sorting systems, *J. Am. Chem. Soc.*, **127**, pp. 15959–15967.

58. Cao, L., Sekutor, M., Zavalij, P. Y., Mlinaric-Majerski, K., Glaser, R. and Isaacs, L. (2014). Cucurbit[7]uril·guest pair with an attomolar dissociation constant, *Angew. Chem. Int. Ed.*, **53**, pp. 988–993.

59. Green, N. M. (1963). Avidin. 1. The use of [^{14}C]biotin for kinetic studies and for assay, *Biochem. J.*, **89**, pp. 585–591.

60. Murray, J., Kim, K., Ogoshi, T., Yao, W. and Gibb, B. C. (2017). The aqueous supramolecular chemistry of cucurbit[*n*]urils, pillar[*n*]arenes and deep-cavity cavitands, *Chem. Soc. Rev.*, **46**, pp. 2479–2496.

61. Kim, H.-J., Heo, J., Jeon, W. S., Lee, E., Kim, J., Sakamoto, S., Yamaguchi, K. and Kim, K. (2001). Selective inclusion of a hetero-guest pair in a molecular host: formation of stable charge-transfer complexes in cucurbit[8]uril, *Angew. Chem. Int. Ed.*, **40**, pp. 1526–1529.

Chapter 3

Fabrication of Macrocycle-Functionalised Nanosensors

This chapter will review and discuss general approaches to fabricating SERS substrates functionalised with macrocyclic molecules. In particular, it will first cover the basic theory and production routes of plasmonic nanoparticles (NP), as key functional building blocks of the SERS substrates. We will focus on bottom-up approaches via wet chemical synthesis, owing to its compatibility with large-scale production and the fact that most of the reported macrocycle-functionalised SERS substrates to date were made via this approach. Secondly, this chapter will discuss general means of attaching macrocycles onto the surface of the plasmonic NPs, which will then allow capturing of target molecules close to the plasmonic hotspots and in some cases simultaneously mediate the formation of the final SERS substrates via self-assembly. Lastly, this chapter will review and discuss strategies for the formation of SERS substrates via self-assembly from plasmonic and macrocyclic building blocks.

3.1 Surface and Interface Chemistry

Different from bulk materials, surface and interface chemistry is particularly important for NPs due to the high surface-area-

Macrocycle-Functionalised Nanosensors
Weng-I Katherine Chio and Tung-Chun Lee

ISBN 978-981-5129-33-5 (Hardcover), 978-1-003-51057-4 (eBook)
www.jennystanford.com

to-volume ratio. In contrast to the saturated inner atoms, the unsaturated surface atoms of NPs have dangling bonds from the incomplete bonding environment, resulting in higher surface energy and reactivity [1]. NPs are thermodynamically unstable and tend to aggregate in the absence of stabilising agents. Ligands can be used to stabilise NPs or functionalise them further for enhanced properties via ligand exchange. A good understanding of surface and interface chemistry at the nanoscale is thus essential to synthesise high-quality plasmonic NPs for sensing applications.

3.1.1 Nucleation of NPs

Synthesis of NPs depends on the formation of a supersaturated solution occurred when the concentration of solute exceeds the solubility limit, leading to a high Gibbs free energy which is then reduced by forming a second phase via solute segregation [2]. The total free energy (ΔG) is the sum of surface free energy (ΔG_s) and volume free energy (ΔG_v) upon formation of a new phase (Fig. 3.1a) and is given by:

$$\Delta G = \Delta\mu_s + \Delta\mu_v \tag{3.1}$$

$$\Delta\mu_s = 4\pi r^2 \gamma \tag{3.2}$$

$$\Delta\mu_v = \frac{4}{3}\pi r^3 \Delta G_v \tag{3.3}$$

where r is the radius of the NP, γ is surface tension, and ΔG_v is Gibbs free energy per unit volume.

ΔG_v is given by [3]:

$$\Delta G_v = \frac{-k_B T \ln(S)}{v} \tag{3.4}$$

where k_B is Boltzmann's constant, T is temperature, S is supersaturation of the solution and v is molar volume.

By combining equations 3.1–3.3, ΔG is thus given by:

$$\Delta G = 4\pi r^2 \gamma + \frac{4}{3}\pi r^3 \Delta G_v \tag{3.5}$$

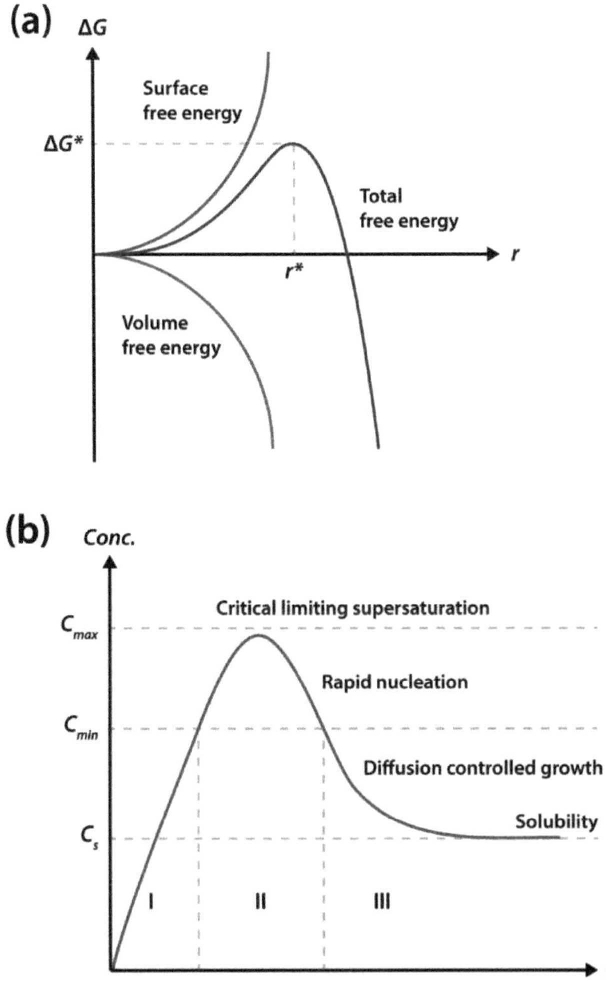

Figure 3.1 Schematic illustration of (a) the change in total Gibbs free energy as a function of radius of the NP and (b) the LaMer model showing different stages of nucleation and growth as a function of time.

To avoid redissolution, the radius of a newly formed NP has to be larger than the critical radius (r^*), the minimum size of a stable NP in solution, which is obtained by differentiating ΔG with respect to r and equating to zero:

$$r^* = -\frac{2\gamma}{\Delta G_v} \tag{3.6}$$

The critical energy (ΔG^*), which is the energy barrier that must be overcome for nucleation to happen, is essentially ΔG at $r^* = r$:

$$\Delta G^* = \frac{16\pi\gamma^3}{3(\Delta G_v)^2} \tag{3.7}$$

According to the LaMer model, a classical theory based on the separation of nucleation and growth, the process of nucleation and growth of NPs can be divided into three stages (Fig. 3.1b) [2–4]. In stage I, there is a rapid increase in the concentration of monomers in the solution until reaching a critical concentration above the solubility limit where supersaturation starts to occur (C_{min}). In stage II, the monomers undergo burst nucleation to form nuclei as the supersaturation is enough to overcome the critical energy. As nucleation proceeds, the concentration of monomers decreases significantly until falling below C_{min} where no more nucleation can take place. In stage III, stable nuclei with sizes greater than the critical radius grow larger by diffusion of monomers till the concentration reaches equilibrium. For monodispersed NPs, the nuclei should be generated at the same time under the same conditions although it also depends on the kinetics of the subsequent growth process.

3.1.2 Growth of NPs

The growth of NPs consists of multi-steps that can be categorised into two mechanisms, diffusion and surface reaction. The former refers to the process of transferring monomers to the growth surface while the latter refers to the process of incorporating adsorbed monomers into the solid [2]. The total flux of monomers (J) passing through a spherical surface within a diffusion layer can be modelled by Fick's first law (Fig. 3.2) [3, 5]:

$$J = 4\pi x^2 D\frac{dC}{dx} \tag{3.8}$$

where x is the radius of the spherical surface, D is the diffusion coefficient, and C is the concentration of monomers at x.

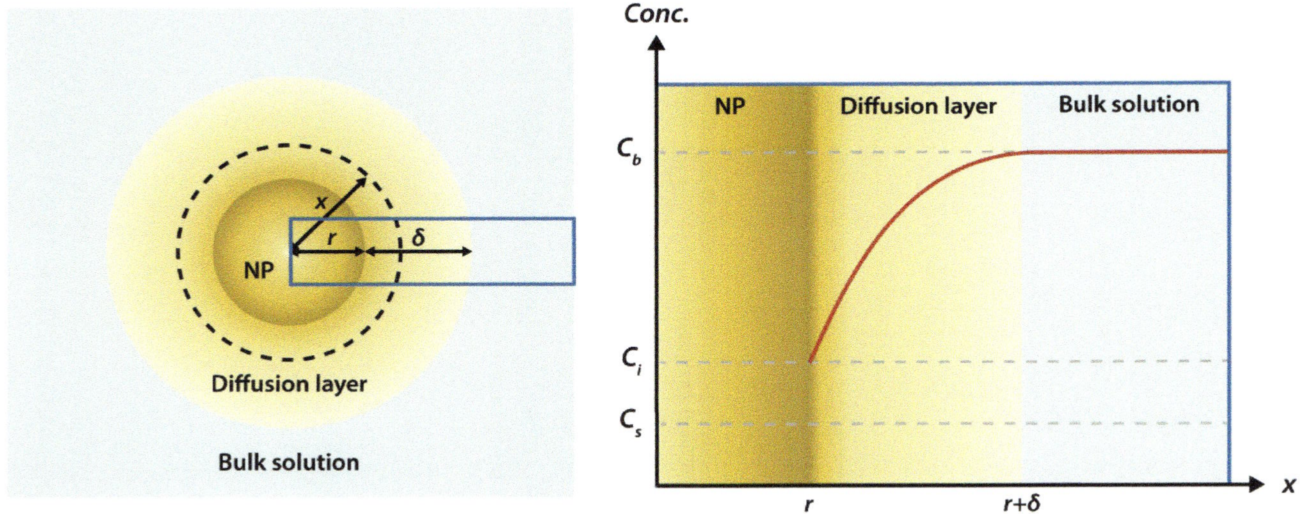

Figure 3.2 Schematic illustrations of the diffusion layer around a NP and the concentration profile of monomers in the diffusion layer.

As J is constant regardless of x for steady-state diffusion, J for a NP in solution can be obtained by integration of $C(x)$ from $r + \delta$ to r:

$$J = \frac{4\pi Dr(r+\delta)}{\delta}(C_b - C_i) \tag{3.9}$$

where r is the radius of the NP, δ is thickness of the diffusion layer, C_b is bulk concentration of monomers, and C_i is concentration of monomers at the solid–liquid interface.

Similarly, the surface reaction is given by:

$$J = 4\pi r^2 k(C_i - C_s) \tag{3.10}$$

where k is the rate constant and C_s is the concentration at the solubility limit.

The relationship between diffusion and surface reaction can be deduced from equations 3.9 and 3.10:

$$\frac{C_i - C_s}{C_b - C_i} = \frac{D}{kr}\left(1 + \frac{r}{\delta}\right) \tag{3.11}$$

3.1.2.1 Diffusion-controlled growth

When the concentration of monomers decreases to the critical concentration C_{min} (Fig. 3.1b), nucleation stops, and the growth of NPs becomes diffusion controlled. As $D \ll kr$, the concentration of monomers at the interface is approximately equal to the concentration at the solubility limit ($C_i \approx C_s$),

$$J = \frac{4\pi Dr(r+\delta)}{\delta}(C_b - C_s) \tag{3.12}$$

When the thickness of the diffusion layer is much greater than the radius of the NP ($\delta \gg r$),

$$J = 4\pi Dr(C_b - C_s) \tag{3.13}$$

As J is related to the rate of change of NP size (dr/dt) by:

$$J = \frac{4\pi r^2}{v}\left(\frac{dr}{dt}\right) \tag{3.14}$$

where v is molar volume.

For diffusion-controlled growth, the rate of change of particle size can be obtained from equations 3.13 and 3.14,

$$\frac{dr}{dt} = \frac{Dv\left(C_b - C_s\right)}{r} \tag{3.15}$$

By solving equation 3.15,

$$r^2 = 2D(C_b - C_s)vt + r_0^{\,2} \tag{3.16}$$

where r_0 is the initial radius of the NP.

The relative radius difference ($\delta r/r$) can be obtained by differentiating equation 3.16 with respect to r and then dividing both sides by r^2,

$$\delta r = \frac{r_0 \delta r_0}{r} \tag{3.17}$$

$$\frac{\delta r}{r} = \frac{r_0 \delta r_0}{r^2} \tag{3.18}$$

$$\frac{\delta r}{r} = \frac{r_0 \delta r_0}{2D\left(C_b - C_s\right)vt + r_0^{\,2}} \tag{3.19}$$

Thus, the relative radius difference between NPs reduces with increasing particle size and time, leading to the formation of monodispersed NPs (Fig. 3.3).

3.1.2.2 Reaction-controlled growth

When diffusion is rapid ($D \gg kr$), the growth of NPs becomes controlled by surface reaction and the bulk concentration of monomers is approximately equal to the concentration of monomers at the interface ($C_b \approx C_i$). The rate of change of NP size in reaction-controlled growth can be deduced from equations 3.10, 3.11, and 3.14,

$$\frac{dr}{dt} = kv\left(C_b - C_s\right) \tag{3.20}$$

This implies the rate of change of NP size is independent of the particle size and time.

By solving the above equation,

$$r = kv(C_b - C_s)t + r_0 \tag{3.21}$$

Differentiating with respect to r and dividing both sides by r,

$$\delta r = \delta r_0 \tag{3.22}$$

$$\frac{\delta r}{r} = \frac{\delta r_0}{r} \tag{3.23}$$

$$\frac{\delta r}{r} = \frac{\delta r_0}{kv(C_b - C_s)t + r_0} \tag{3.24}$$

The relative radius difference between NPs reduces with increasing particle size and time, leading to the formation of monodispersed NPs (Fig. 3.3). This is known as polynuclear growth, one of the two mechanisms of reaction-controlled growth. It is a rapid process in which the next layer begins to grow prior to the completion of the previous layer.

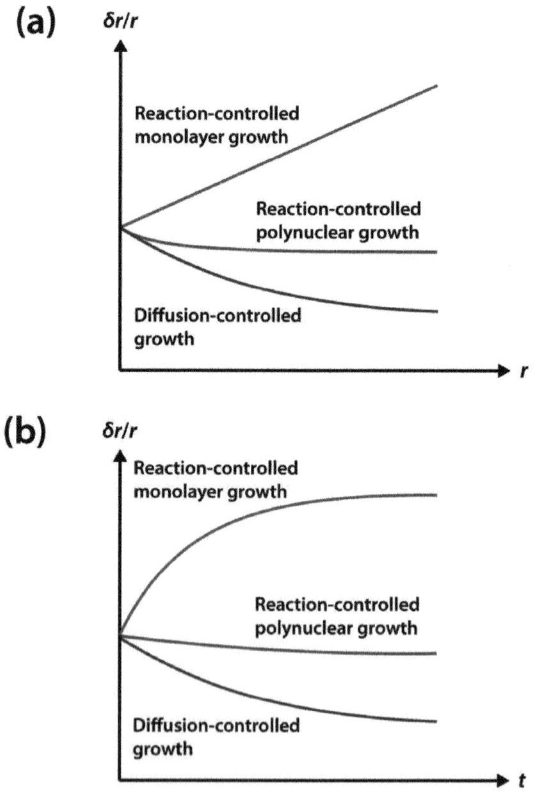

Figure 3.3 Schematic illustrations of radius difference as a function of (a) radius of the NP and (b) growth time for the three different growth mechanisms [2].

The second mechanism of reaction-controlled growth is a layer-by-layer process known as monolayer growth, a special case in which the growth of the previous layer is completed before the following commences.

The rate of change of the NP size is given by:

$$\frac{dr}{dt} = k^* r^2 \tag{3.25}$$

where k^* is the rate constant for mononuclear growth.

By solving the above equation and differentiating with respect to r,

$$\frac{1}{r} = \frac{1}{r_0} - k^* t \tag{3.26}$$

$$\frac{\delta r}{r} = \frac{r \delta r_0}{r_0^2} \tag{3.27}$$

$$\frac{\delta r}{r} = \frac{\delta r_0}{1 - r_0 kt} \tag{3.28}$$

For monolayer growth, the relative size difference increases with larger particle size and longer growth time, which implies it is not favourable for the growth of monodispersed NPs.

Note that the three mechanisms become predominant at different particle sizes under favourable conditions [2]. Monolayer and polynuclear growth dominate at small and medium particle sizes whereas diffusion-controlled growth becomes predominant at larger particle sizes. Nonetheless, diffusion-controlled growth is essential for growth of monodispersed NPs by homogeneous nucleation.

3.1.3 Ostwald Ripening

When the size of colloidal particles decreases, the increased solubility and surface energy of the particles make them more unstable in solution. The phase instability of the particles can be described by the Gibbs-Thomson equation [6]:

$$C_s = C_b \exp\left(\frac{2\gamma v}{r k_B T}\right) \tag{3.29}$$

where C_s is the concentration at solubility limit, C_b is bulk concentration, γ is surface energy, v is molar volume, r is the radius of the particle, k_B is Boltzmann's constant, and T is temperature.

The small colloidal particles then redissolve to allow further growth of the larger particles in a process known as Ostwald ripening. Note that it is also possible to have the small particles growing at the expense of the larger particles in a process called digestive ripening, which is the opposite of Ostwald ripening.

3.2 Synthesis of Nanoparticles

While the very first application of plasmonic NPs dates back to the 4th century (see Section 1.2), no synthetic method had been reported until Michael Faraday synthesised Au NPs via phosphorus reduction of $AuCl_3$ solution in a two-phase CS_2-water mixture in 1857 [7]. Since then, many different techniques, which can be classified into two categories, the top-down and bottom-up approaches, have been developed to prepare plasmonic NPs. The former refers to the fabrication of NPs via gradual size reduction from bulk materials while the latter refers to the synthesis of NPs from atoms, molecules, and clusters. Top-down approaches such as attrition milling and lithography are relatively cheap and suitable for mass production whereas it is challenging to process and maintain a good homogeneity at nanoscale. On the other side, bottom-up approaches such as colloidal self-assembly, chemical vapour deposition, and physical vapour deposition are more commonly used for the synthesis of plasmonic NPs, thanks to their narrow size distribution and high uniformity. Despite the huge number of reports on the synthetic methods of plasmonic NPs, the Turkevich method developed in 1951 and the Brust-Schiffrin method developed in 1994 remainthe most popular methods for Au NP synthesis to date. Subsequently, a wide variety of methods have also been developed for the synthesis of non-spherical plasmonic NPs and semi-conductor NPs (i.e., quantum dots, QDs), with the former often involving the use of surfactants that limit the functionalisation by macrocycles to Type IV.

3.2.1 The Turkevich Method

The breakthrough in the synthesis of Au NPs did not take place until a century after Faraday's report. In 1951, Turkevich and co-workers developed a similar approach to synthesise Au NPs with a typical size of 20 nm via sodium citrate reduction of chloroauric acid ($HAuCl_4$) in an aqueous solution at 100°C [8]. During the reaction, Au^{3+} from the $HAuCl_4$ is reduced to Au^0, the solution becomes supersaturated in which the Au NPs can be precipitated. The colour of the solution changes from pale yellow through colourless and dark blue to ruby red. Later in 1973, Frens refined the Turkevich method to synthesise Au NPs with sizes ranging from 16 to 147 nm by varying the ratio of sodium citrate to $HAuCl_4$, which results in polydispersity when the size of NPs is greater than 20 nm [9]. It is postulated that the stabilisation mechanism involves large particles that shrink over time to form stable Au colloids [10].

Continued efforts in the modification of the Turkevich method, including temperature, pH, ratio of reactants, and order of addition, and understanding of the mechanisms have been made in recent decades. For instance, a combination of sodium citrate and tannic acid can be used to reduce $HAuCl_4$ in the synthesis of sub–20 nm Au NPs at a lower temperature of 60°C [11, 12] while ascorbic acid alone can be used as a reducing agent to synthesise Au NPs with sizes ranging from 20 to 100 nm in the expense of broad size distribution at ≥50 nm at room temperature [13]. To account for the dark blue intermediate present in the course of Au NP synthesis via the conventional Turkevich method, it is suggested that the growth mechanism involves self-assembly of 5 nm Au clusters to form a network of nanowires which subsequently fragments into spherical NPs [14]. On the other hand, the Au NP synthesis is proposed to occur via a multi-step mechanism, in which Au^{3+} is first reduced to Au^+ before disproportionates to Au^0 meanwhile citrate is oxidised to dicarboxy acetone, leading to nucleation and growth of Au NPs [15]. In 2010, in situ small angle X-ray scattering (SAXS) and X-ray absorption near edge spectroscopy (XANES) were coupled to monitor the growth process of Au NPs, which is proposed as a multi-step mechanism involving nuclei coalescence and monomer attachment [16]. It is suggested that the coalescence process is critical for the size distribution of the Au NPs.

In addition, the pH of the solution depends on the concentration of sodium citrate used in the synthesis, which in turn determines the size distribution of Au NPs [17]. For instance, monodispersed Au NPs ranging from 20 to 40 nm can be synthesised by simply varying pH of the solution. As sodium citrate is a weak reducing agent, the Au NPs synthesised via the Turkevich method are usually larger than 15 nm. Monodispersed Au NPs of 9–15 nm can be synthesised by reversing the order of reagent additions, i.e., injecting $HAuCl_4$ into the sodium citrate solution [18]. An exhaustive list of modifications to the Turkevich method can be found in a recent publication [19]. The Au NPs synthesised via citrate reduction are suitable for further functionalisation by macrocycles as the citrate-capping layer is only loosely bound to the surface of the Au NPs.

3.2.1.1 Seeded growth

In 2011, Bastus and co-workers developed a kinetically controlled seeded growth method to synthesise monodispersed Au NPs with controllable sizes in the range of 8.4–180.5 nm, simply by varying temperature, pH, and concentration of Au precursor to seeds [20]. A lower temperature is used to inhibit secondary nucleation in the reaction while the control of pH and concentration improves the monodispersity of the Au NPs by controlling the reactivity of Au species and shifting the systems towards the focusing conditions. The citrate-capped Au NPs with sizes ≥40 nm prepared by the seeded growth method are particularly useful for the SERS nanosensor (see Section 3.4). The detailed procedures for the preparation of such particles are as follows: 99.5 mL of a 0.25 mM $HAuCl_4$ solution is refluxed in a 250 mL three-necked round-bottomed flask equipped with a condenser at 90°C for 15 min. 0.5 mL of a 500 mM sodium citrate solution is then injected into the flask and refluxed until the colour of the solution turns ruby red. The synthesised Au seeds are cooled to 70°C, and 0.67 mL of a 25 mM $HAuCl_4$ solution and 0.67 mL of a 60 mM sodium citrate solution are sequentially injected with a time interval of 2 min. This sequential injection process is then repeated to gradually increase the size of Au NPs to 40 nm (Fig. 3.4). A sample can be extracted after 30 or 60 min for characterisation by UV-Vis spectroscopy, dynamic light scattering (DLS), and transmission electron microscopy (TEM) (see Section 3.4). The average size of the Au NPs can be calculated by measuring 200 particles from the TEM images.

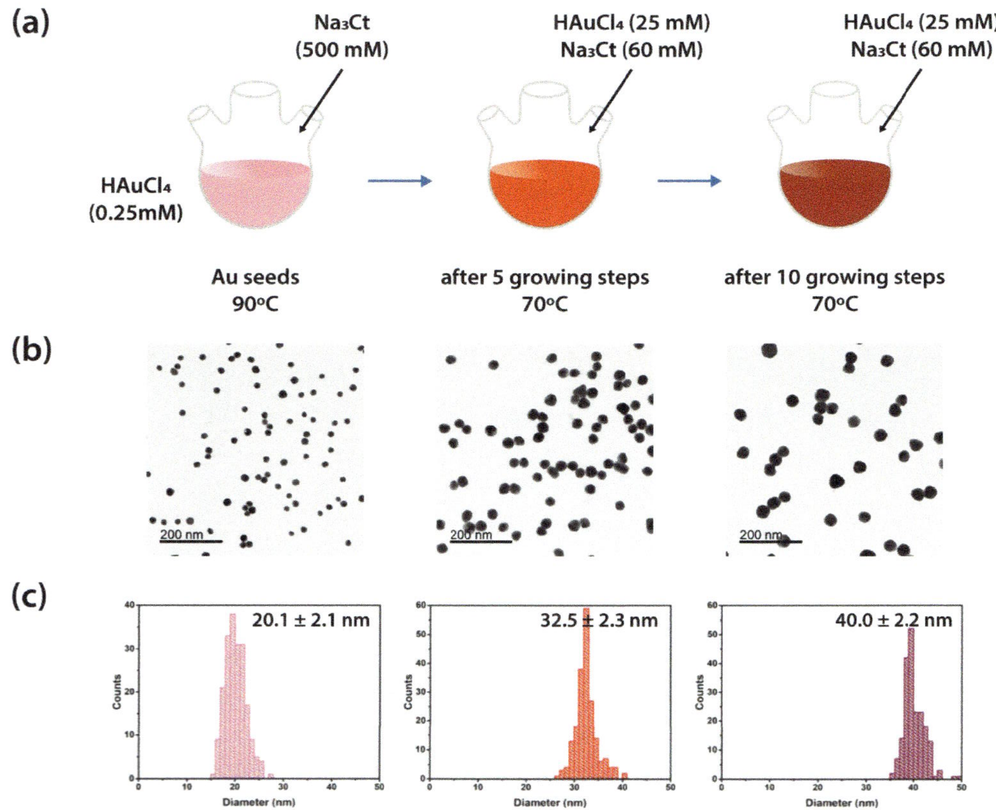

Figure 3.4 (a) Schematic illustration of seeded growth of Au NPs. (b) TEM images and (c) histograms showing the different sizes of Au NPs produced by the kinetically controlled seeded growth method.

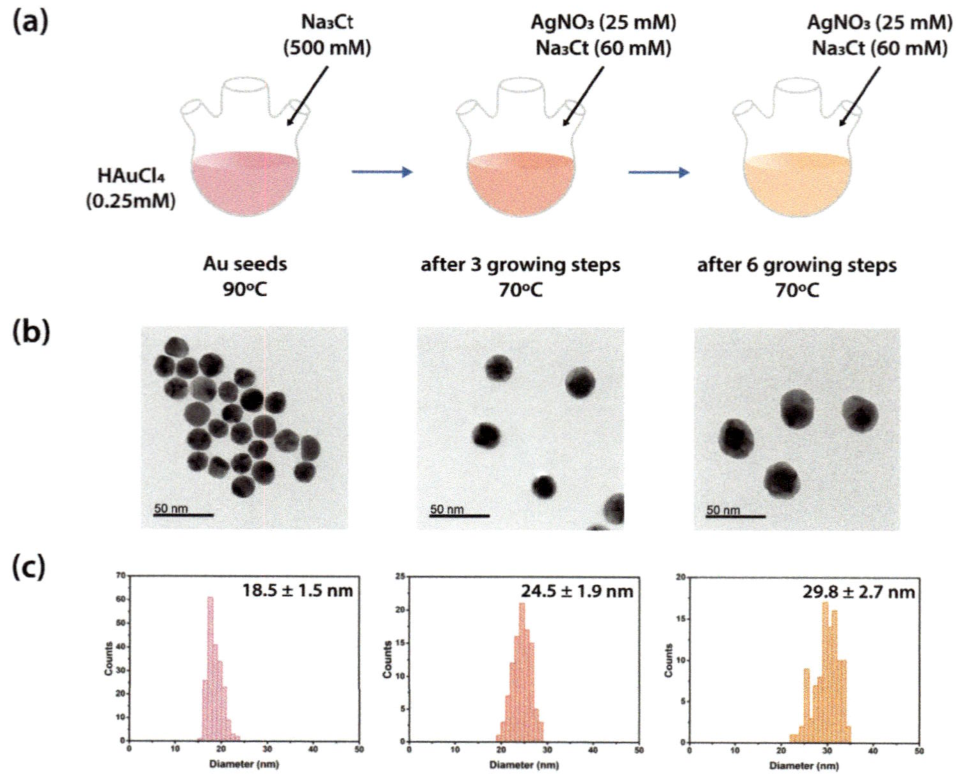

Figure 3.5 (a) Schematic illustration of seeded growth of Au-Ag core-shell NPs. (b) TEM images and (c) histograms showing the different sizes of Au-Ag core-shell NPs produced by the kinetically controlled seeded growth method.

In addition, Au-Ag core-shell NPs can be synthesised using the seeded growth method for enhanced sensitivity of the SERS nanosensor, due to the intrinsically stronger SERS signals and the strong dependence of SERS enhancement on distance (see Section 4.7). In particular, 0.67 mL of a 25 mM $AgNO_3$ solution and 0.67 mL of a 60 mM sodium citrate solution are sequentially injected into the synthesised Au seeds at 70°C with a time interval of 2 min. This sequential injection process is then repeated to grow core-shell NPs with increasing Ag shell thickness (Fig. 3.5). The colour of the solution gradually changes from ruby red to orange.

The seeded growth method is preferred over the Lee-Meisel method [21] in which citrate-capped Ag NPs are synthesised via sodium citrate reduction of $AgNO_3$ as the conventional synthetic method results in polydispersed NPs over a range of sizes between 60 and 200 nm.

3.2.2 The Brust-Schiffrin Method

While the Turkevich method has been widely used to synthesise Au NPs, the synthesis of sub-10 nm Au NPs remained an open challenge until Brust and co-workers developed a one-step method to synthesise Au NPs with stabilising thiol ligands in 1994 [22]. The Brust-Schiffrin method utilises a two-phase (water-toluene) system to allow the surface reaction to take place during nucleation and growth. During the synthesis, $AuCl_4^-$ is transferred from the aqueous phase to the organic phase with tetraoctylammonium bromide (TOAB, $(C_8H_{17})_4BrN$) acting as a phase-transfer reagent, before being reduced by sodium borohydride ($NaBH_4$) in the presence of dodecanethiol ($C_{12}H_{25}SH$) to form a deep brown solution (Fig. 3.6a). The thiol ligands self-assemble on the surface of Au NPs which prevents agglomeration of Au NPs due to the strong interaction between Au and thiol. This method allows the synthesis of thiol-functionalised Au NPs ranging from 1 to 3 nm in a mixture of cuboctahedral and icosahedral shapes (Fig. 3.6b). The same group further optimised the method by using p-mercatophenol as a stabilising agent for Au NPs in a single-phase system without the need for the phase-transfer agent, [23] paving the way for the synthesis of different thiol-stabilised Au NPs. Although thiols are strongly attached to the surface of the Au NPs and prohibit formation of Type III CB-NPs, this and a similar method can be used to make Type IV CB-NPs (see Section 1.3).

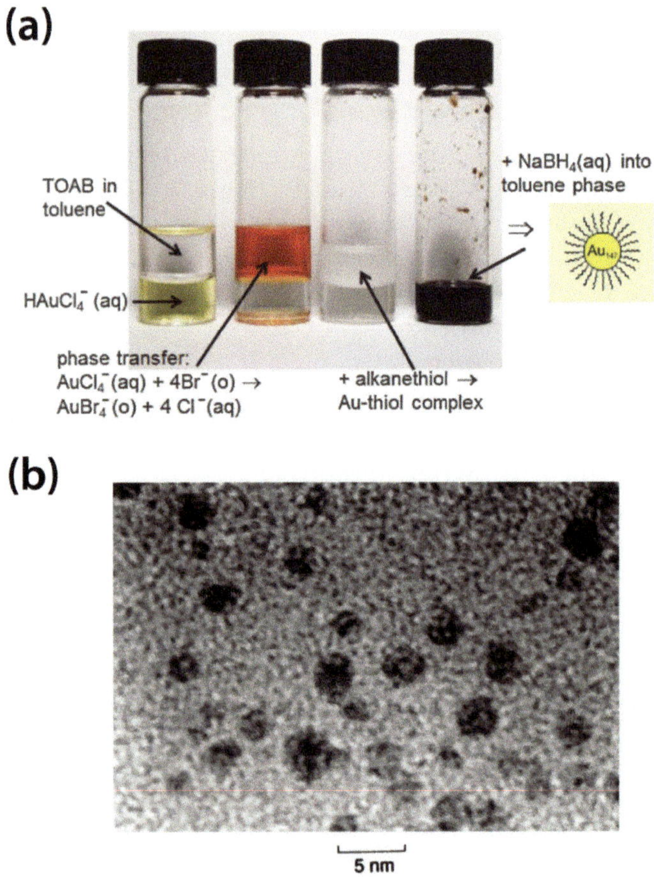

Figure 3.6 (a) Schematic illustration of the Brust-Schffrin method and (b) TEM image of the thiol-stabilised Au NPs synthesised via the Brust-Schffrin method [22, 24].

3.2.3 Polyol Synthesis

In 1989, Fievet, Lagier, and Figlarz developed a method to synthesise microparticles (e.g., Co, Ni, and Cu) using polysol (multivalent alcohols) which was subsequently extended to the synthesis of other metal NPs including Ag NPs [25]. Spherical Ag NPs with sizes of 5–140 nm can be synthesised via $NaBH_4$ reduction of $AgNO_3$ in various polysol such as ethylene glycol (EG), 1,2-propanediol, 1,4-butanediol, and 1,5-pentanediol at room or elevated temperature [26]. Besides, the polyol approach can also be used for the synthesis of Ag nanocubes (NCs) and Ag nanocages. For instance, NCs with an edge length of sub 20 nm can be synthesised using diethylene glycol (DEG), which has a higher viscosity and lower reducing power than EG, to achieve a purity of >95% [27]. In this method, the Ag precursor, silver trifluoroacetate (CH_3COOAg), was reduced by DEG in the presence of sodium hydrosulfide (NaSH), HCl, and poly(vinyl pyrrolidone) (PVP). The edge length of Au NCs can be tuned from 18 to 32 nm by varying reaction time (Fig. 3.7). Note that the Ag NCs can also be used to prepare Au nanocages via galvanic replacement reaction with $HAuCl_4$ solution for nanosensing.

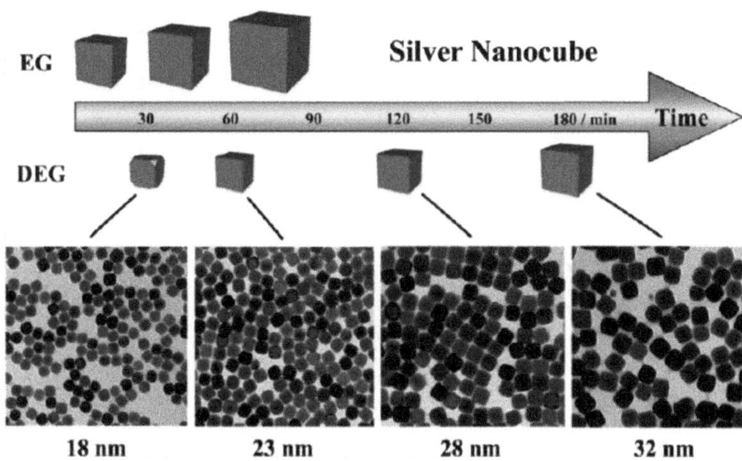

Figure 3.7 Schematic illustration and TEM images of Ag NCs synthesised using the polyol method [27].

3.2.4 Slow Growth and Slow Chemical Etching

In 2013, Lee and co-workers developed a method to synthesise ultrasmooth and highly spherical monocrystalline Au NPs with sizes up to 200 nm for reproducible plasmonics [28]. This method involves a cyclic process of slow growth to produce single-crystalline polyhedral NPs and slow chemical etching to remove edges and vertices of the NPs. During synthesis, the $HAuCl_4$ solution is reduced by EG in the presence of poly(diallyl dimethyl ammonium chloride) and phosphoric acid at 195°C. The colour of the solution changes from yellow, through colourless and purple, to finally brown, indicating the formation of octahedral Au NPs. Then, $HAuCl_4$ solution is added to the synthesised NPs as an oxidizing agent after cooling. The colour of the solution changes from brown to pink while the shape of the Au NPs changes from octahedra to truncated octahedra, quasi-spheres, and finally spheres over 20 h (Fig. 3.8). The slow etching process introduces isotropic surface tension and thus favours the formation of spherical NPs under quasi-static conditions. Due to the presence of relatively stronger ligands, this method is only limited to the formation of Type II and IV CB-NPs.

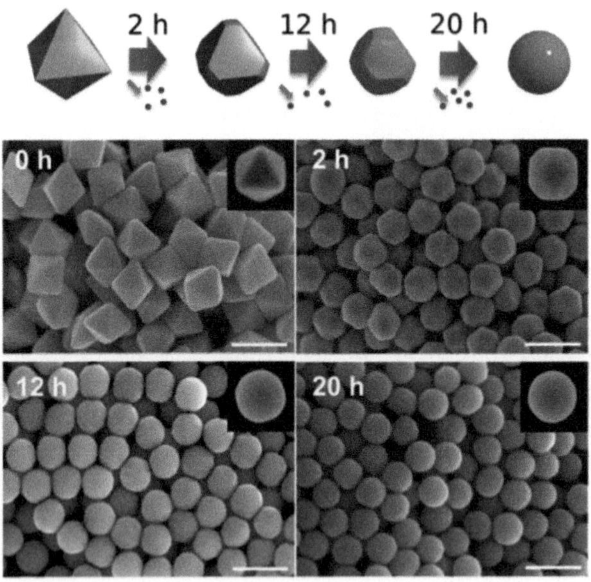

Figure 3.8 Schematic illustration and SEM images of Au NPs showing shape evolution from octahedra to spheres [28].

3.2.5 Other Synthetic Methods for Metal NPs

Apart from the aforementioned synthetic methods, there are many other methods that can be used to synthesise plasmonic NPs. For instance, Jana and Peng developed single-phase approaches to synthesise nearly monodispersed NPs of 1–15 nm with relatively weaker ligands or surfactants than thiol in 2003 [29]. A combination of tetrabutylammonium borohydride (TBAB) and hydrazine in toluene is used to reduce the metal precursors, $AuCl_3$, silver acetate ($Ag(CH_3COO)$), and anhydrous copper(II) acetate ($Cu(CH_3COO)_2$), to form the respective plasmonic NPs (Fig. 3.9), of which the surface can be subsequently modified via ligand exchange. On the other hand, Zheng and co-workers developed a facile method to synthesise single-crystal Au NPs in the range of 5–150 nm [30]. In particular, Au seeds can be synthesised by adding $NaBH_4$ to $HAuCl_4$ with cetyltrimethylammonium bromide (CTAB) as a capping agent and subsequently grown to a larger size with cetyltrimethylammonium chloride (CTAC) as a stabilising agent and ascorbic acid as reducing agent. Similar to the Brust-Schiffrin method, these synthetic methods involving ligands stronger than the macrocycles are also only limited to the formation of Type IV CB-NPs.

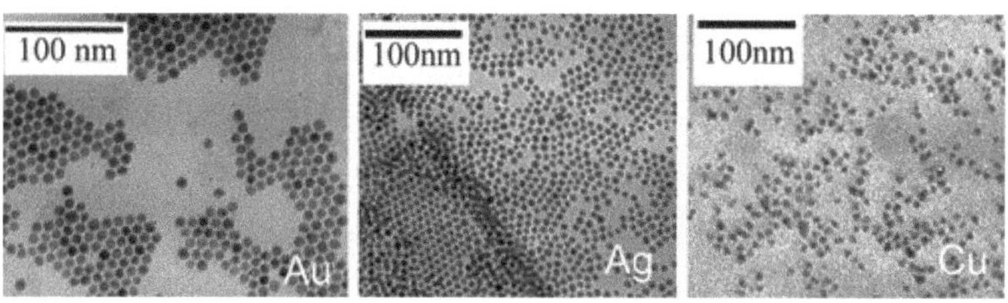

Figure 3.9 TEM images of Au, Ag, and Cu NPs synthesised with TBAB and hydrazine as reducing agents [29].

3.2.6 Synthesis of Quantum Dots

Recently, the direct binding between CB and mercaptopropionic acid-coated quantum dots (MPA-coated QDs) has been reported, showing potential applications in trace chemical detection by nanosensors. In 2008, Zou and co-workers developed a rapid method to synthesise highly luminescent CdTe QDs in an aqueous environment, which are suitable for further attachments of macrocycles [31]. In particular, tellurium powder and $NaBH_4$ are dissolved in water and the mixture is heated under nitrogen flow at 80°C for 30 min until the colour of the solution changes to deep red, indicating the formation of NaHTe. On the other side, $CdCl_2$ and MPA are dissolved in water and basified with NaOH solution to pH 12. Then, NaHTe is added to the Cd/MPA solution at room temperature before heating to 100°C under reflux. MPA-coated CdTe QDs of different emissions from green to near-infrared can be produced by varying the growth time (Fig. 3.10).

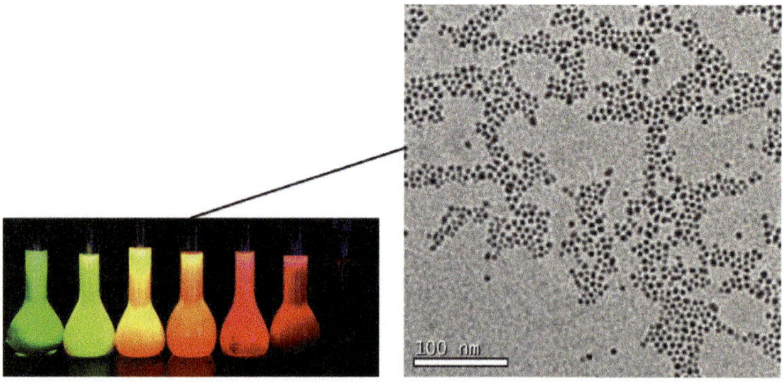

Figure 3.10 Photograph of MPA-coated CdTe QDs with a growth time of 1, 5, 10, 20, 30, 45, 60, and 90 min under UV radiation and the corresponding TEM image of the QDs with a growth time of 30 min [31].

3.3 Attachment of Macrocycles onto the Surface of Nanoparticles

As bare plasmonic NPs tend to aggregate to lower their surface energy, it is necessary to precisely control their aggregations to ensure the formation of reproducible SERS substrates. Functional ligands and linkers such as citrate, alkylthiols, carboxylates, amines, DNA, and biotin-streptavidin often fail to produce a controllable and consistent inter-particle spacing between NPs or provide access to plasmonic hotspots, resulting in irreproducible plasmonics and hence are not desirable for use as SERS nanosensors [2]. On the other side, macrocycles such as cucurbiturils (CBs) can be used to define a precise inter-particle spacing between NPs and form plasmonic hotspots. There are three general approaches to attach macrocycles onto the surface of plasmonic NPs, including in situ formation, ligand displacement, and host–guest complexation.

3.3.1 In Situ Formation of Plasmonic NPs

In situ formation of plasmonic NPs is a very facile approach to attach macrocycles; however, the size, polydispersity and surface chemistry of the resultant plasmonic NPs cannot be precisely controlled. Thus, for better control over the final morphology and the surface chemistry of the plasmonic nanostructures, it is beneficial to decouple the synthesis of plasmonic NPs, which can be performed using classical techniques, and the attachment of macrocyclic molecules.

While the idea of encapsulating Au NPs within the CB cavity was suggested by Corma and co-workers in 2007, the use of CBs as stabilisers for Au NPs was first reported by Lee and Scheman three years later. They synthesised Au NPs via $NaBH_4$ reduction of $HAuCl_4$ in the presence of CB5 and showed the feasibility of forming CB5-Au NPs nanoaggregates both before and after the reduction. The macrocycles can be singly capped or doubly capped by the plasmonic NPs with a precise spacing of 0.91 nm between adjacent Au NPs, which is effectively the height of CB molecules, depending on the ratio of CB:Au NPs (Fig. 3.11). Note that the ratio of single-capped

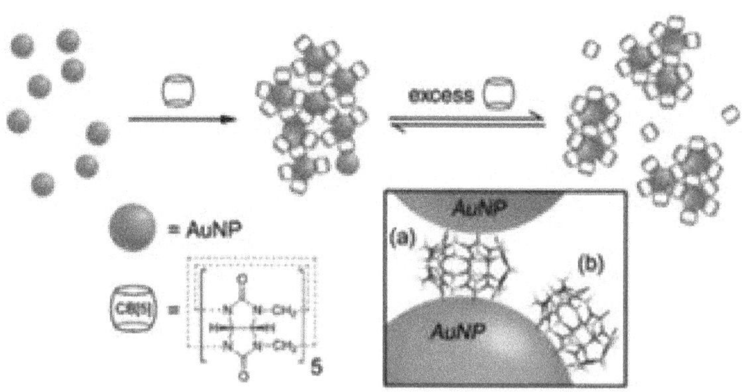

Figure 3.11 Schematic illustration of attachment of CB5 onto the surface of Au NPs via their electron-rich carbonyl portals. Inset: (a) Doubly and (b) singly-capping of CB5 by Au NPs [32].

CB5 increases with the CB5:Au NP ratio. In this approach, CBs are attached onto the surface of Au NPs via their electron-rich carbonyl portals as in the case of a flat Au surface. The carbonyl portals donate electrons to the surface of Au NPs and weaken the C=O bond. The attachment of CBs onto Au NPs is mainly driven by the enthalpic gain from the interaction between carbonyl and Au, and the entropic gain from the release of surrounding water molecules. The macrocycle-NP nanoaggregates formed via this approach have led to subsequent applications in plasmonic nanosensing.

3.3.2 Ligand Displacement

Ligand displacement is a simple but effective approach to attach macrocycles onto plasmonic NPs. Typically, plasmonic NPs can be synthesised via classical synthetic methods discussed in Section 3.2. Meanwhile, macrocyclic molecules can be functionalised with a thiol group that can bind strongly to Au NPs or QDs. For instance, per-thiolated CDs (α-, β-, or γ-) can be attached to Au NPs via reduction of $AuCl_4^-$ by $NaBH_4$ in a DMSO solution containing the respective CDs to form water-soluble CD-modified Au NPs with sizes of 2–7 nm [33]. In addition, the thiol-modified CB6-QD complex can be prepared by Type II binding (Fig. 3.12) [34]. CB6 modified with approximately two thiol groups linked on the sides can form host–guest complexes with N-(6-aminohexyl)-4-nitrobenzenamine (NBzA) and bind onto the surface of a QD.

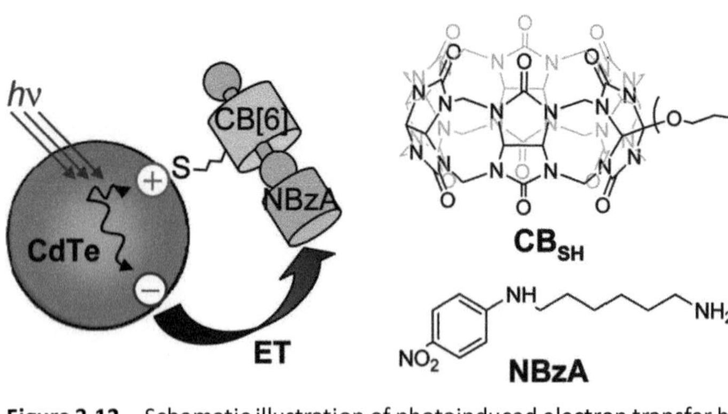

Figure 3.12 Schematic illustration of photoinduced electron transfer between CdTe QD and NBzA encapsulated in thiolated CB6 [34].

Attachment of macrocycles can be performed routinely via simple mixing of plasmonic NPs protected by weak or medium strength ligands and thiolated macrocycles. Notably, unfunctionalised, native CBs were found to exhibit binding interactions with Au or Ag surfaces. They can displace very weak ligands (e.g., Cl^-) and weak ligands (e.g., citrate) on Au NPs to form CB-Au NPs [35, 36]. The ligand displacement occurs upon simple mixing of aqueous solutions of the corresponding components and is typically complete within seconds. Interestingly, the initially formed CB-NPs can spontaneously assemble to form plasmonic nanojunctions and subsequently larger NP clusters with controllable morphology.

3.3.3 Host–Guest Complexation

Alternatively, the macrocycles can also be attached onto the NP surface via indirect host–guest interactions. In this scenario, guest molecules will be tethered to a strong ligand such as thiol and amine. Thiolated guest molecules can then be attached onto the NP surface via ligand displacement to form guest-functionalised NPs. The surface-bound guest moieties can then be used to capture macrocyclic molecules via host–guest complexation. For instance, the Scherman group prepared Type IV CB8-Au NPs in an aqueous solution via the formation of 1:1:1 ternary host–guest complexes of CB8. Methyl viologen (MV^{2+}), an electron-deficient first guest, is used to functionalise Au NPs in the presence of naphthol-containing copolymers, an electron-rich second guest (Fig. 3.12) [37]. The same group also developed a one-step fabrication of microcapsules held together by 1:1:1 ternary host–guest complexes of CB8 with methyl viologen naphthol-containing copolymers at a liquid–liquid interface. The microcapsules could potentially be used for nanosensing of the encapsulated molecules [38].

3.4 Fabrication of Macrocycle-Functionalised SERS Substrates

To develop high-performance SERS nanosensors, it is essential to understand how the enhancement and reproducibility of the signals are affected by the features of the SERS substrates. Though single-molecule detection is of considerable interest in the research community, other aspects including reproducibility, uniformity, ease-of-fabrication, stability, and cost-effectiveness are more critical to sensing applications. Herein, various CB-NP SERS substrates based on the state of matters, i.e., solution and solid, are presented in the following subsections.

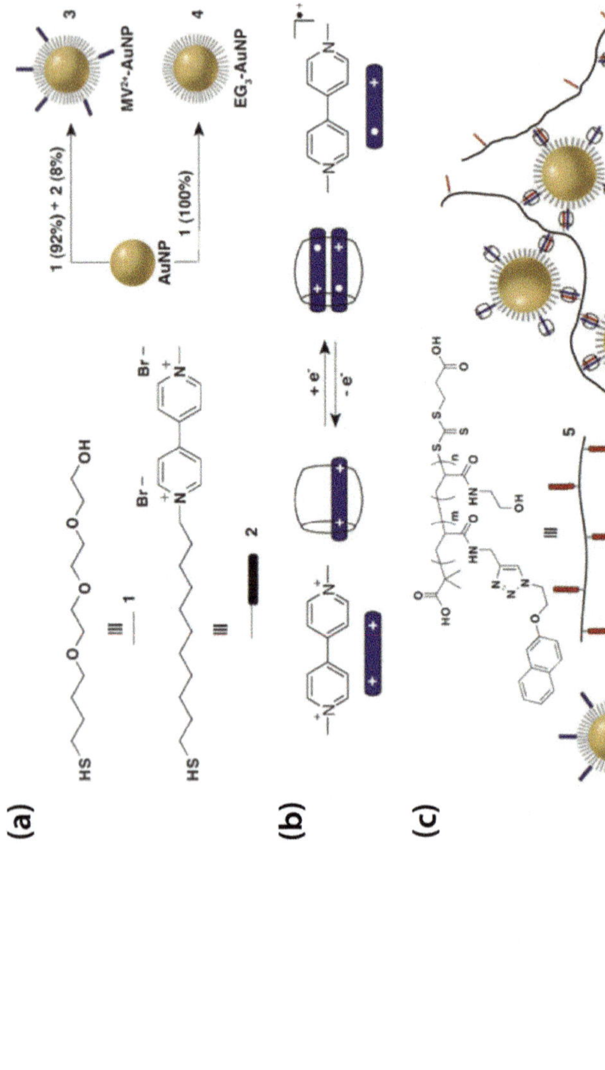

Figure 3.13 Schematic illustration of (a) MV^{2+}-Au NP and EG_3-Au NP, (b) a 1:2 CB8-MV^{2+} host–guest complex, and (c) the indirect CB8-Au NP interaction. Note that EG_3 indicates tri(ethylene glycol) [37].

3.4.1 Solution-Based Substrates

While colloidal NPs are the most commonly used solution-based SERS substrates, unaggregated spherical NPs suffer from low SERS signal enhancement due to their high symmetry. The SERS signals of colloidal NPs can be enhanced via controlled aggregation to form a 3D network of plasmonic hotspots, which can be triggered by adding CBs. The ability of CBs to simultaneously mediate the formation of precise plasmonic nanojunctions and recognise small molecules via host–guest complexation is particularly important for the detection of analyte targets with low affinities to the metal surface.

The solution-based CB-NP SERS substrates can be simply produced by mixing the CB and colloidal NP solution. The aggregation process of CB-NPs can be classified into two kinetic growth regimes, diffusion-limited colloidal growth (DLCA) and reaction-limited colloidal growth (RLCA), depending on the ratio of CB: NP [36]. For instance, the aggregation follows the DLCA regime at a high CB5:Au NP ratio (80:1) but follows the RLCA regime at a low CB5:Au NP ratio (60:1). In the DLCA regime, the LSPR peak of Au NPs at 525 nm is slightly red-shifted with decreased intensity upon aggregation. Also, a strong secondary peak appears at 650 nm which is then red-shifted with increased intensity (Fig. 3.14a). In the RLCA regime, the aggregation of Au NPs follows a similar trend as in the DLCA regime but at a much slower speed, showing the reproducibility of the plasmonic SERS substrates (Fig. 3.14b). At the high CB concentration, open and elongated chain-like structures are first formed as the colliding particles cannot reach the centre of the nanoclusters but instead coagulate with the outermost particles, consistent with the DLCA regime. The chain-like structures then join together as quasi-fractal networks (Fig. 3.14c). On the other side, slow formation of compact nanoclusters can be observed at the low CB concentration, following the RLCA regime (Fig. 3.14d). In addition, the CB-NP SERS substrates can also be fabricated in situ, where the CB host–guest complexes are first formed, followed by adding that to NPs to trigger aggregation (see Section 5.2.3). Interestingly, it is possible to use a combination of different CB homologues as SERS substrates. For example, CB5 can be used to align Au nanorods (NRs). Meanwhile, CB7 can be used to form host–guest complexes with R6G. CBs are preferentially bound to the (111) end facet of the NRs as the (100)

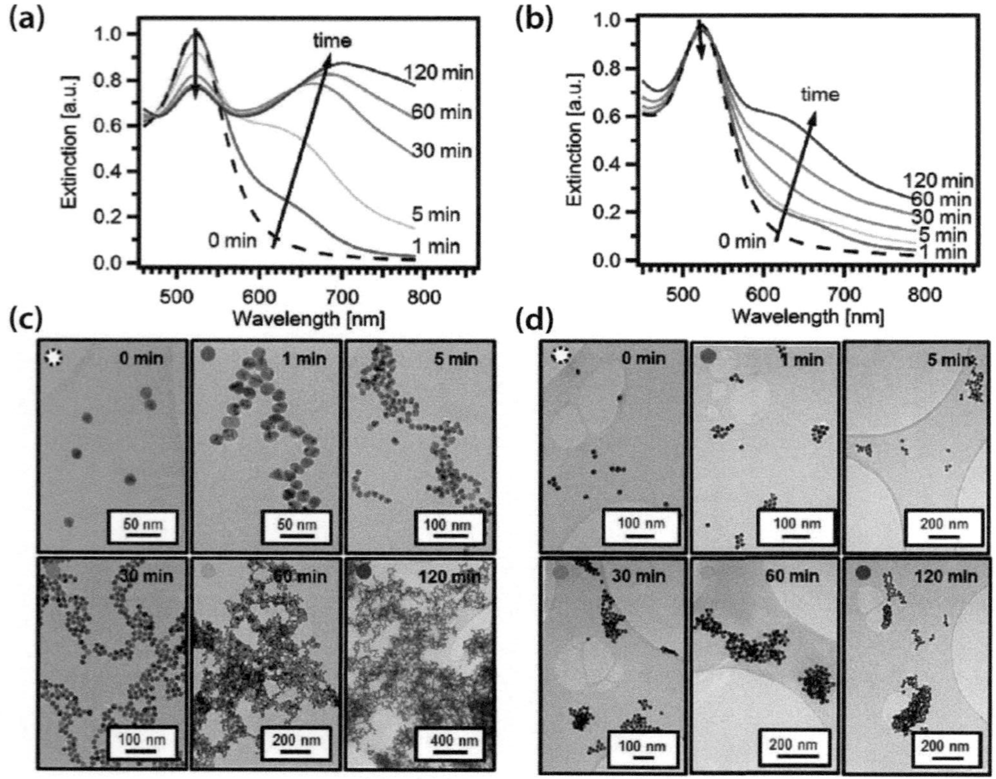

Figure 3.14 Optical extinction of CB5-Au NP nanoaggregates with a ratio of (a) 80:1 and (b) 60:1. The corresponding TEM images of the CB5-Au NP nanoaggregates with a ratio of (c) 80:1 and (d) 60:1 [36].

longitudinal facet is strongly capped by CTAB and cannot be easily displaced. Note that the solubility of less soluble CB homologues such as CB8 can be enhanced by adding 10 mM HCl (see Section 7.2.3).

3.4.2 Solid-Based Substrates

Apart from aqueous CB-NP nanoaggregates, solid-based SERS substrates exploiting the plasmonic nanojunctions defined by CBs have also been used. For example, macrocycle-functionalised SERS substrates can be simply fabricated by dropping CB*n* solution onto commercial nanostructured substrates (e.g., Klarite) or drying aqueous CB-Au NP nanoaggregates on a glass slide, Si wafer, or Au film. In addition, NPs on mirror geometry (NPoM), i.e., Au NPs on Au film sandwiched by CB, are also a popular type of CB-NP SERS substrates (Fig. 3.15) [39]. An adhesion layer (e.g., Cr, Ti) is deposited on a Si wafer, followed by the deposition of the Au film using techniques such as electron-beam evaporation or physical vapour deposition. The Au film is immersed into a pre-mixed CB7 host–guest solution for 24 h to form a monolayer. Au NPs are then drop-casted onto the substrates to produce NPoM. Alternatively, the CB-NP SERS substrates can be fabricated using a layer-by-layer approach. The Au film is immersed into the CB7 solution for 24 h to form a monolayer of CB on the Au surface. The substrate is then spin-coated with Au NP solution before immersing in CB7 for an hour. The process is repeated several times to build a nanoporous structure on the Au surface. The reproducibility of the SERS signals using multi-layered substrates is relatively higher than the monolayered substrates at the expense of uniformity issues due to aggregations of NPs.

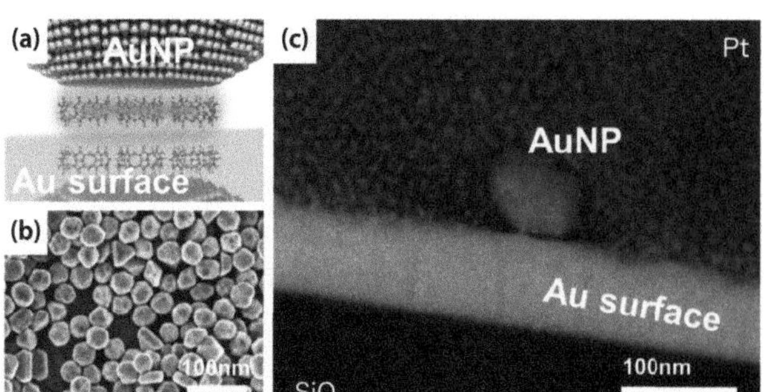

Figure 3.15 (a) Schematic illustration of NPoM, (b) SEM image of 100 nm Au NPs, and (c) dark-field STEM image of NPoM [39].

3.5 Characterisation of Macrocycle-Functionalised Nanostructures

In addition to synthesis and fabrication, the ability to characterise nanomaterials plays an important role in the development of high performance nanosensors. Materials characterisation techniques such as optical spectroscopies and electron microscopies enable us to visualise the macrocycle-functionalised NPs and to better understand their properties for the designated applications. UV-Vis spectroscopy, dynamic light scattering, Fourier transform infrared (FT-IR) spectroscopy, transmission electron microscopy, and scanning electron microscopy are examples of the most commonly used techniques for nanoparticles before and after macrocycle functionalisation.

3.5.1 UV-Visible Spectroscopy

UV-Vis spectroscopy is one of the earliest, yet simplest methods used to illustrate the structural difference between as-synthesised and macrocycle-functionalised NPs (see Section 4.2 for fundamentals of UV-Vis spectroscopy). For instance, Au NPs synthesised in the absence of CB show a characteristic LSPR peak at 520 nm in the UV-Vis spectrum (Fig. 3.16, trace (a)) [32]. When CB5 is present in the synthesis with a molar ratio of 0.1, the LSPR peak is red-shifted and a secondary LSPR peak is observed at 620 nm, thus indicating the formation of larger 3D nanoaggregates and 1D chain-like self-assembled nanostructures, respectively (b). As the molar ratio of CB5:Au NP doubles to 0.2, the 520 nm LSPR peak is red-shifted further whereas the 620 nm disappears, illustrating the presence of 3D nanoaggregates but the absence of the 1D nanostructures (c). Interestingly, as the molar ratio of CB:Au NP increases further to 0.5 and 1, the LSPR peak at 520 nm approaches that of the Au NP sample in the absence of CB (d and e). This implies the size of the 3D nanoaggregates decreases but the ratio of single-capped CB increases.

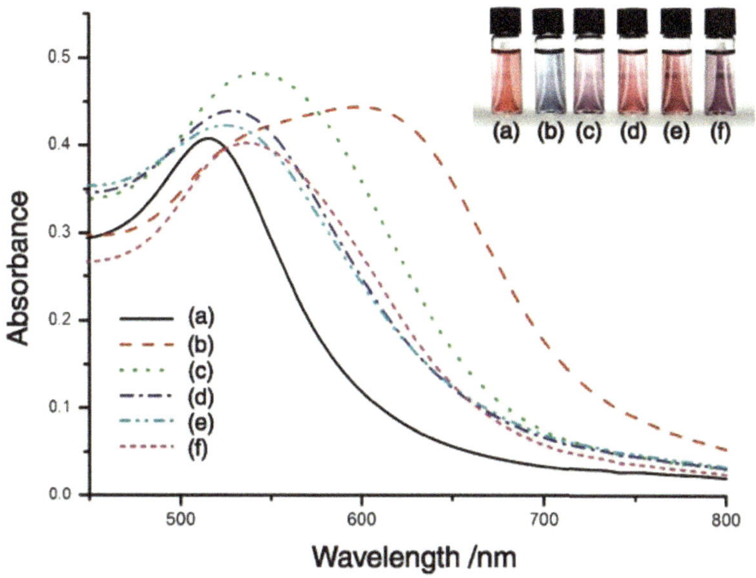

Figure 3.16 UV-Vis spectra and photographs of CB5-Au NPs with molar ratios of (a) 0, (b) 0.1, (c) 0.2, (d) 0.5, and (e) 1 [32].

3.5.2 Dynamic Light Scattering Technique

DLS techniques can be used to measure the size of macrocycle-functionalised NPs in solution besides UV-Vis spectroscopy (see Section 4.5 for the fundamentals of DLS). For instance, the hydrodynamic size of Au NPs synthesised via $NaBH_4$ reduction of $HAuCl_4$ was measured to be ~12 nm using DLS (Fig. 3.17(a)). In the presence of CB5 with a molar ratio of 0.1, the average size of the CB5-Au NP nanoaggregates increases to 323 nm, indicating the formation of chain-like nanostructures (b). By increasing the CB5:Au NP molar ratio to 0.2, 0.5, and 1, the hydrodynamic size of the nanoaggregates decreases to 70, 10, and 10 nm, respectively (c–e). Note that the monotonic decrease in nanoaggregate size is consistent with the UV-Vis results (see Section 3.4.1).

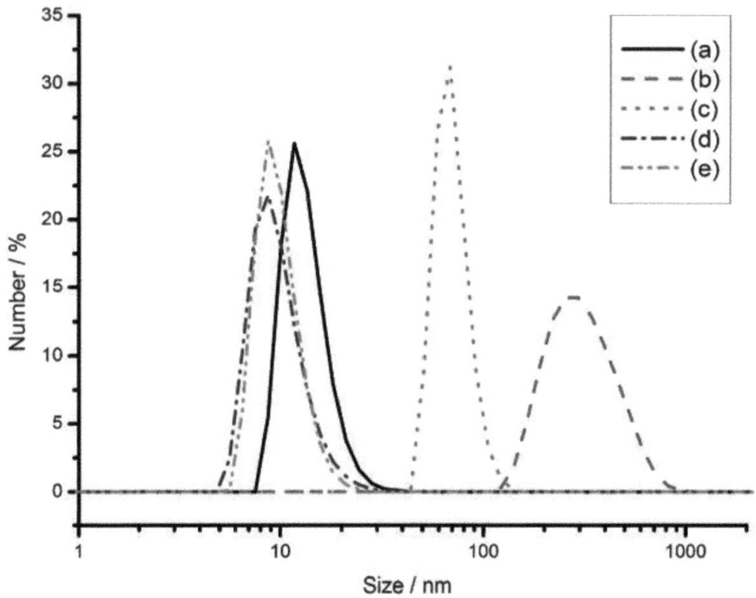

Figure 3.17 Hydrodynamic size of CB5-Au NPs with molar ratios of (a) 0, (b) 0.1, (c) 0.2, (d) 0.5, and (e) 1 [32].

3.5.3 Fourier Transform Infrared Spectroscopy

FT-IR can be used to study the capping of macrocycles by NPs (see Section 4.4 for fundamentals of IR spectroscopy). For example, CB5 has a characteristic absorption peak at 1733 cm^{-1}, which can be attributed to carbonyl stretching at the portal (Fig. 3.18). Interestingly, the absorption peak is red-shifted by ~15 cm^{-1} in the CB5-Au NP nanoaggregates, indicating that capping by Au NPs is via the carbonyl portal with electrons being donated to the Au NP surface and the C=O being weakened in the self-assembly.

3.5.4 Nuclear Magnetic Resonance Spectroscopy

Nuclear magnetic resonance (NMR) spectroscopy is a non-destructive analytical technique used to identify complex molecular structures and investigate the molecular interactions in solution or solid-state samples. The underlying principle is based on the magnetic properties of specific atomic nuclei in the sample that possess an odd number of protons and/or neutrons (e.g., ^{1}H and ^{13}C) in response to an external magnetic field. In NMR spectroscopy, a glass tube containing the sample is placed inside a strong magnet (Fig. 3.19). The specific atomic nuclei absorb and re-emit electromagnetic radiation in the radiofrequency range when their spins are aligned with the magnetic field. The RF signals are then picked up by the detector and converted into a NMR spectrum, giving information about the electronic environment known as chemical shifts.

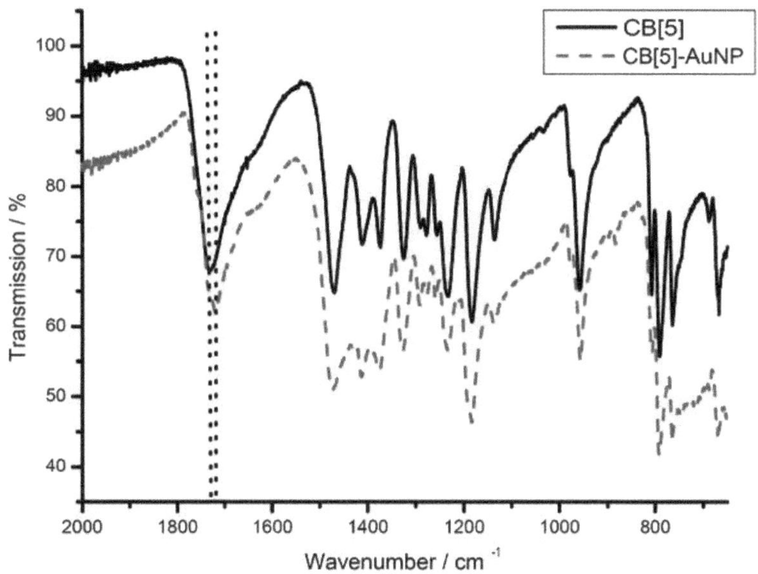

Figure 3.18 FT-IR spectra of CB5 and CB5-Au NP nanoaggregates [32].

NMR spectroscopy can be used to study the capping of macrocycles by QDs. For instance, when CB7 is added to the MPA-coated CdTe QD solution, the molecular environments of the stabiliser (MPA) at 2.3–3.0 ppm, attributed to CH_2-CH_2, can be observed in the proton (^{1}H) NMR spectra (Fig. 3.20). The NMR signals of MPA increase with increasing concentration of CB, indicating that MPA on the QD surface is displaced by CB. In addition, the CB7 proton signals at 4.10, 5.40, and 5.65 ppm are downfield shifted and heavily broadened due to the interaction between CB7 and QDs. Note that the CB7 signals broaden further with increasing CB7 concentration, as more CB7 molecules are driven to the surface of QDs, thus increasing the size of the CB7-QD nanoaggregates and slowing diffusion on NMR timescale.

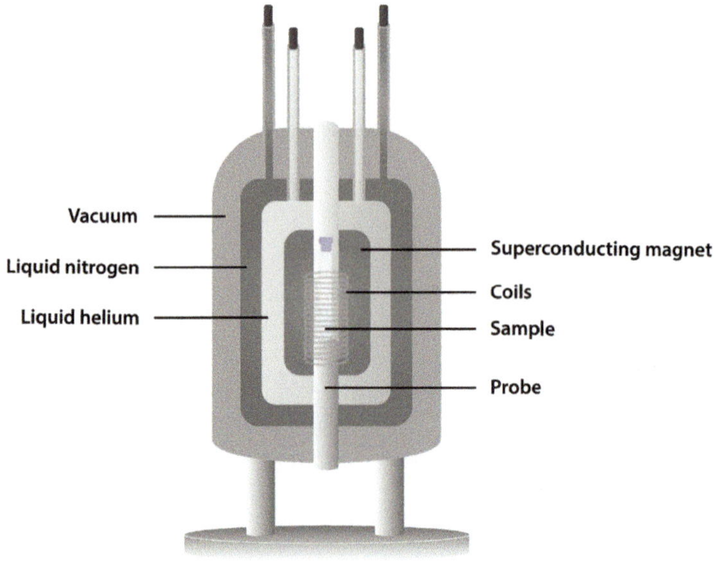

Figure 3.19 Schematic illustration of a NMR spectrometer.

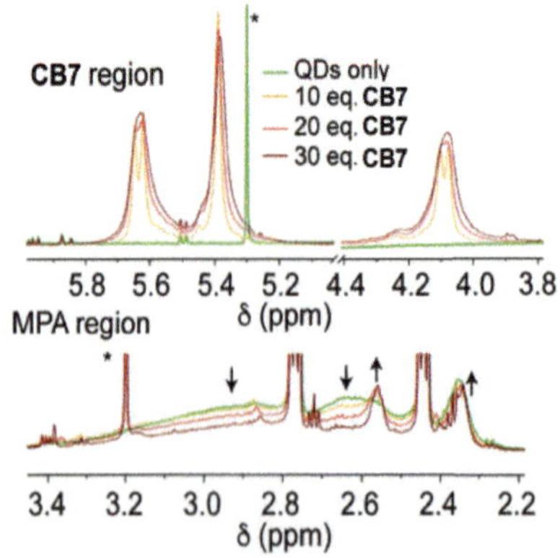

Figure 3.20 ^{1}H NMR spectra of MPA-coated QDs and CB7-QDs in D_2O. Changes with increasing concentrations of CB7 are indicated by arrows while the solvent residues are marked by the asterisk (*) [40].

3.5.5 Electron Microscopies

Electron microscopy is a powerful technique used to characterise the structural features and compositions of the samples with a very high spatial resolution, making it particularly useful for nanotechnology. Unlike optical microscopies, electron microscopies utilise a beam of electrons as the source of illumination in order to overcome the diffraction limit of light and allow visualisation of nanostructures. There are two main types of electron microscopy: transmission electron microscopy (TEM) and scanning electron microscopy (SEM) (Fig. 3.21). The main difference between them is that TEM uses transmitted electrons, but SEM uses reflected or scattered electrons to generate the image. Note that the sample preparation process may lead to artefacts and the samples may also be damaged in both TEM and SEM.

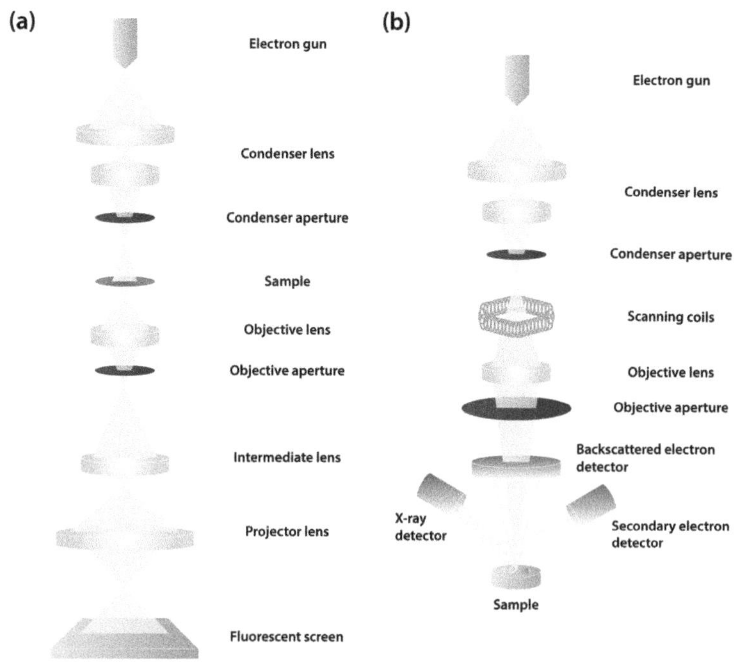

Figure 3.21 Schematic illustrations of (a) TEM and (b) SEM.

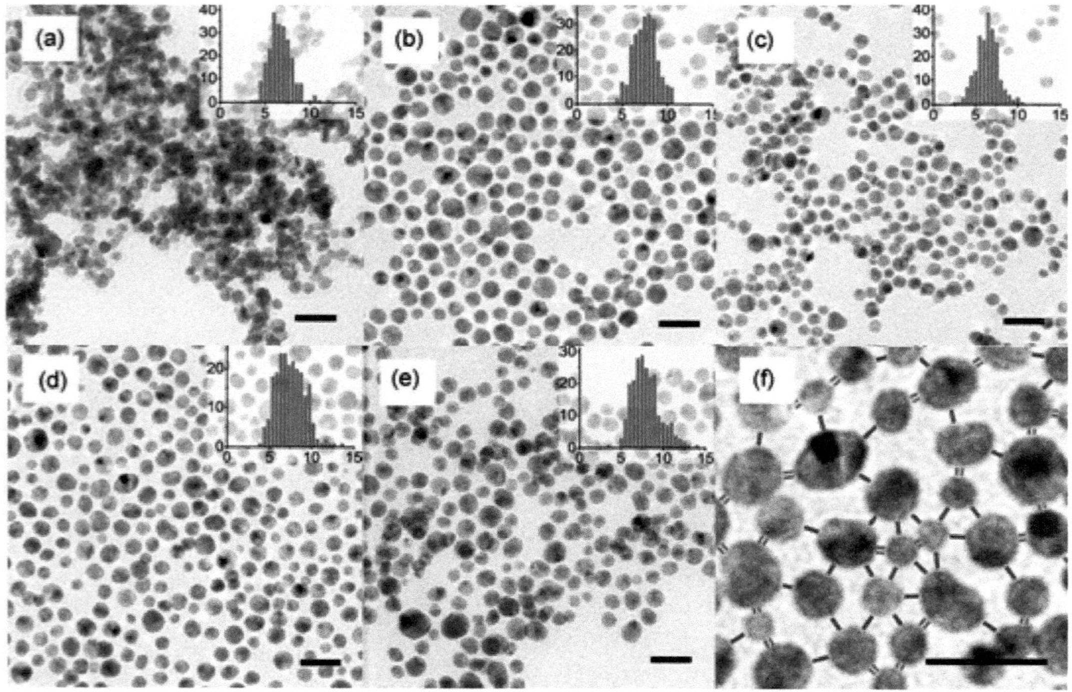

Figure 3.22 TEM images of (a) metastable Au NPs, (b-e) CBn-Au NPs nanoaggregates with n = 5, 6, 7, 8 accordingly, and (f) zoom-in of (d) showing two singly capped (represented by "–") and a doubly capped (represented by "=") CB7. The scale bar indicates 20 nm [35].

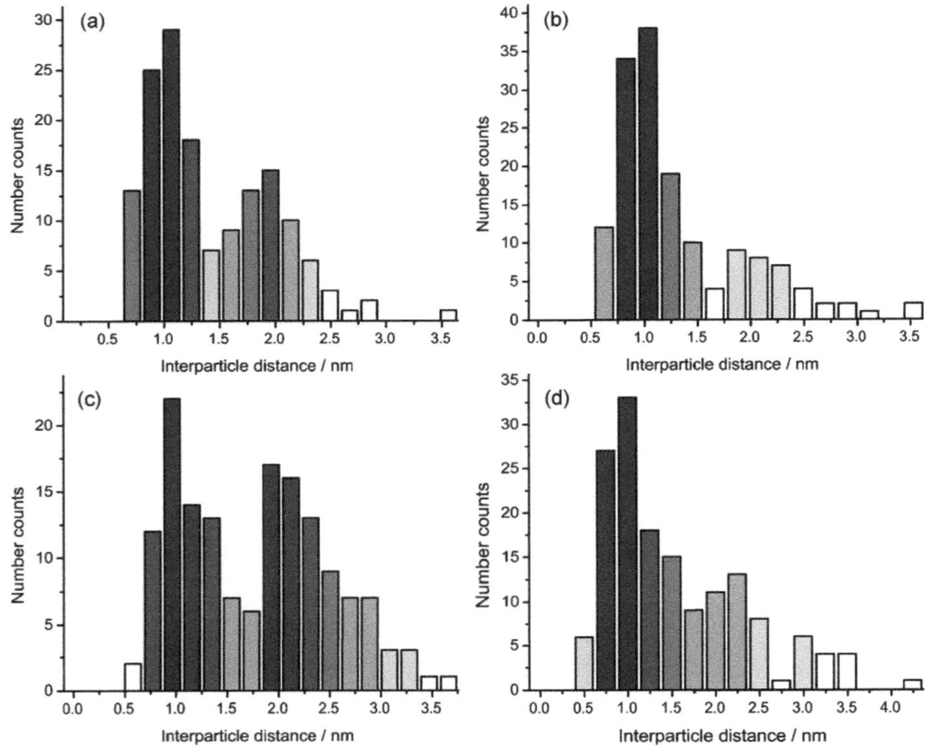

Figure 3.23 Histograms showing the interparticle spacing between Au NPs and their nearest neighbours in CB*n*-Au NPs nanoaggregates with *n* = (a) 5, (b) 6, (c) 7, and (d) 8 [35].

3.5.5.1 Transmission electron microscopy

TEM consists of an electron source, a series of electrostatic and electromagnetic lenses and apertures, and an imaging system (Fig. 3.21a). In TEM, a beam of high-energy electrons is transmitted through an ultra-thin (<100 nm) specimen, and the electrons interact with the atoms in the sample to form an image, which is then magnified and focused onto a fluorescent screen or charge-coupled device. This technique can be used to study the structure of macrocycle-functionalised NPs such as morphology and crystal structure. The specimens are prepared by simply adding drops of the colloidal NP solution onto a Cu TEM grid.

The size and shape of the Au NPs and CB*n*-Au NPs (n = 5–8) can be visualised on TEM and the mean diameters can be calculated by sampling 200 particles (Fig. 3.22) [35]. From the TEM images, the regular spacings between the Au NPs in the presence of CB show the macrocycles are effectively capped by the Au NPs. Note that the interparticle spacing of the adjacent Au NPs is around ~1.0 nm, which is consistent with the height of the CB*n* molecules (Fig. 3.23). This further verifies the FT-IR results that CB*n* interacts with the NPs via their electron-rich carbonyl portals. Interestingly, there is a secondary peak at ~2.0 nm for the CB*n*-Au NP nanoaggregates, which can be attributed to the presence of two singly capped CBs close to each other on the TEM grid. Note that the ratios of the areas under the main and secondary peaks in the histograms are higher for CB6-Au NPs and CB8-Au NPs than CB5-Au NPs and CB7-Au NPs, meaning that they contain a larger portion of the doubly capped CBs, probably due to their relatively low solubilities and concentrations in solution.

3.5.5.2 Scanning electron microscopy

SEM consists of an electron source, a series of lenses and apertures, and detectors as in TEM meanwhile also containing a set of scanning coils (Fig. 3.21b). In SEM, a high energy electron beam is rastered across the surface and interacts with the sample to generate different signals including backscattered electrons (BSE), secondary electrons (SE), and X-rays which are then collected by the detectors to create an image. This technique is suitable for probing surface morphology, elemental composition, and microstructures of materials such as complex macrocycle-metal nanostructured substrates. For instance, the formation of CB7-Au NPs nano-islands or nanoaggregates on an Au film by alternating dip-coating of CB7 and spin-coating of 40 nm Au NPs can be verified using SEM (Fig. 3.24). Note that the electron beam in SEM can heat up the macrocycle-functionalised substrates and darken the focused area quickly.

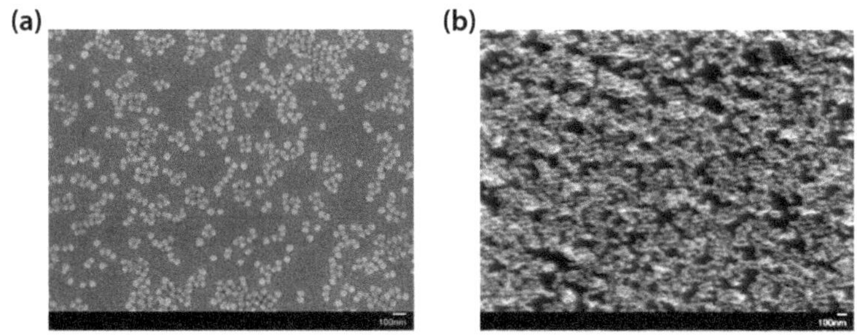

Figure 3.24 SEM images of layer-by-layer nanostructures of Au-CB7-Au NP fabricated by dip- and spin-coating after (a) 5 cycles and (b) 20 cycles.

References

1. Xu, L., Liang, H.-W., Yang, Y. and Yu. S.-H. (2018). Stability and reactivity: positive and negative aspects for nanoparticle processing, *Chem. Rev.*, **118**, pp. 3209–3250.
2. Cao, G. and Wang, Y. (2010). *Nanostructures and Nanomaterials: Synthesis, Properties and Applications*, 2nd ed. (World Scientific, Singapore).
3. Thanh, N. T. K., Maclean, N. and Mahiddine, S. (2014). Mechanisms of nucleation and growth of nanoparticles in solution, *Chem. Rev.*, **114**, pp. 7610–7630.
4. LaMer, V. K. and Dinegar, R. H. (1950). Theory, production and mechanism of formation of monodispersed hydrosols, *J. Am. Chem. Soc.*, **72**, pp. 4847–4854.
5. Sugimoto, T. (1987). Preparation of monodispersed colloidal particles, *Adv. Colloid Interface Sci.*, **28**, pp. 65–108.
6. Gentry, S. T., Kendra, S. F. and Bezpalko, M. W. (2011). Ostward ripening in metallic nanoparticles: stochastic kinetics, *J. Phys. Chem. C.*, **115**, pp. 12736–12741.

7. Faraday, M. (1857). X. The Bakerian lecture – experimental relations of gold (and other metals) to light, *Philos. Trans. Royal Soc.*, **147**, pp. 145–181.
8. Turkevich, J., Stevenson, P. C., Hillier, J. (1951). A study of the nucleation and growth processes in the synthesis of colloidal gold, *Discuss Faraday Soc.*, **11**, pp. 55–75.
9. Frens, G. (1973). Controlled nucleation for the regulation of the particle size in monodisperse gold suspensions, *Nat. Phys. Sci.*, **241**, pp. 20–22.
10. Chow, M. K. and Zukoski, C. F. (1994). Gold sol formation mechanisms: role of colloidal stability, *J. Colloid Interface Sci.*, **165**, pp. 97–109.
11. Muhlpfordt, H. (1982). The preparation of colloidal gold particles using tannic acid as an additional reducing agent, *Experientia*, **38**, pp. 1127–1128.
12. Slot, J. W. and Geuze, H. J. (1985). A new method of preparing gold probes for multiple-labeling cytochemistry, *Eur. J. Cell Biol.*, **38**, pp. 87–93.
13. Kimling, J., Maier, M., Okenve, B., Kotaidis, V., Ballot, H. and Plech, A. (2006). Turkevich method for gold nanoparticle synthesis revisited, *J. Phys. Chem. B*, **110**, pp. 15700–15707.
14. Pong, B.-K., Elim, H. I., Chong, J.-X., Ji, W., Trout, B. L. and Lee, J.-Y. (2007). New insights on the nanoparticle growth mechanism in the citrate reduction of gold (III) salt: formation of the Au nanowire intermediate and its nonlinear optical properties, *J. Phys. Chem. C*, **111**, pp. 6281–6287.
15. Kumar, S., Gandhi, K. S. and Kumar, R. (2007). Modelling of formation of gold nanoparticles by citrate method, *Ind. Eng. Chem. Res.*, **46**, pp. 3128–3136.
16. Polte, J., Ahner, T. T., Delissen, F., Sokolov, S., Emmerling, F., Thunemann, A. F. and Kraehnert, R. (2010). Mechanism of gold nanoparticle formation in the classical citrate synthesis method derived from coupled in situ XANES and SAXS evaluation, *J. Am. Chem. Soc.*, **132**, pp. 1296–1301.
17. Ji, X., Song, X., Li, J., Bai, Y., Yang, W. and Peng, X. (2007). Size control of gold nanocrystals in citrate reduction: the third role of citrate, *J. Am. Chem. Soc.*, **129**, pp. 13939–13948.
18. Ojea-Jimenez, I., Bastus, N. G. and Puntes, V. (2011). Influence of the sequence of the reagents addition in the citrate-mediated synthesis of gold nanoparticles, *J. Phys. Chem. C*, **115**, pp. 15752–15757.

19. Daruich De Souza, C., Ribeiro Nogueira, B. and Rostelato, M. E. C. M. (2019). Review of the methodologies used in the synthesis gold nanoparticles by chemical reduction, *J. Alloys Compd.*, **798**, pp. 714–740.

20. Bastus, N. G., Comenge, J. and Puntes, V. (2011). Kinetically controlled seeded growth synthesis of citrate-stabilized gold nanoparticles for up to 200 nm: size focusing versus Ostwald ripening, *Langmuir*, **27**, pp. 11098–11105.

21. Lee, P. C. and Meisel, D. (1982). Adsorption and surface-enhanced Raman of dyes on silver and gold sols, *J. Phys. Chem.*, **86**, pp. 3391–3395.

22. Brust, M., Walker, M., Bethell, D., Schiffrin, D. J. and Whyman, R. (1994). Synthesis of thiol-derivatised gold nanoparticles in a two-phase liquid-liquid system, *J. Chem. Soc., Chem. Commun.*, **1994**, pp. 801–802.

23. Brust, M., Fink, J., Bethell, D., Schiffrin, D. J. and Kiely, C. (1995). Synthesis and reactions of functionalised gold nanoparticles, *J. Chem. Soc., Chem. Commun.*, **1995**, pp. 1655–1656.

24. Liz-Marzan, L. M. (2013). Gold nanoparticle research before and after the Brust-Schiffrin method, *Chem. Commun.*, **49**, pp. 16–18.

25. Dong, H., Chen, Y.-C. and Feldmann, C. (2015). Polyol synthesis of nanoparticles: status and options regarding metals, oxides, chalcogenides, and non-metal elements, *Green Chem.*, **17**, pp. 4107–4132.

26. Park, K. H., Im, S. H. and Park, O. O. (2011). The size control of silver nanocrystals with different polyols and its application to low-reflection coating materials, *Nanotechnology*, **22**, 045602.

27. Wang, Y., Zheng, Y., Huang, C. Z. and Xia, Y. (2013). Synthesis of Ag nanocubes 18–32 nm in edge length: the effects of polyol on reduction kinetics, size control and reproducibility, *J. Am. Chem. Soc.*, **135**, pp. 1941–1951.

28. Lee, Y.-J., Schade, N. B., Sun, L., Fan, J. A., Bae, D. R., Mariscal, M. M., Lee, G., Capasso, F., Sacanna, S., Manoharan, V. and Yi, G.-R. (2013). Ultrasmooth, highly spherical monocrystalline gold particles for precision plasmonics, *ACS Nano.*, **7**, pp. 11064–11070.

29. Jana, N. R. and Peng, X. (2003). Single-phase and gram-scale routes toward nearly monodisperse Au and other noble metal nanocrystals, *J. Am. Chem. Soc.*, **125**, pp. 14280–14281.

30. Zheng, Y., Zhong, X., Li, Z. and Xia, Y. (2014). Successive, seed-mediated growth for the synthesis of single-crystal gold nanospheres with

uniform diameters controlled in the range of 5–150 nm, *Part. Part. Syst. Charact.*, **31**, pp. 266–273.

31. Zou, L., Gu, Z., Zhang, N., Zhang, Y., Fang, Z., Zhu, W. and Zhong, X. (2008). Ultrafast synthesis of high luminescent green- to near infrared-emitting CdTe nanocrystals in aqueous phase, *J. Mater. Chem.*, **18**, pp. 2807–2815.
32. Lee, T.-C. and Scherman, O. A. (2010). Formation of dynamic aggregates in water by cucurbit[5]uril capped with gold nanoparticles, *Chem. Commun.*, **46**, pp. 2438–2440.
33. Liu, J., Ong, W., Roman, E., Lynn, M. J. and Kaifer, A. E. (2000). Cyclodextrin-modified gold nanospheres, *Langmuir*, **16**, pp. 3000–3002.
34. Cui, S.-C., Tachikawa, T., Fujitsuka, M. and Majima, T. (2011). Photoinduced electron transfer in a quantum dot-cucurbituril supramolecular complex, *J. Phys. Chem. C*, **115**, pp. 1824–1830.
35. Lee, T.-C., Scherman, O. A. (2012). A facile synthesis of dynamic supramolecular aggregates of cucurbit[*n*]uril (n=5–8) capped with gold nanoparticles in aqueous media, *Eur. J. Chem.*, **18**, pp. 1628–1633.
36. Taylor, R. W.; Lee, T.-C., Scherman, O. A., Esteban, R., Aizpurua, J., Huang, F. M., Baumberg, J. J. and Mahajan, S. (2011). Precise subnanometer plasmonic junctions for SERS within gold nanoparticle assemblies using cucurbit[n]uril "glue", *ACS Nano*, **5**, pp. 3878–3887.
37. Coulston, R. J., Jones S. T., Lee, T.-C., Appel, E. A and Scherman, O. A. (2011). Supramolecular gold nanoparticle-polymer composites formed in water with cucurbit[8]uril, *Chem. Commun.*, **47**, pp. 164–166.
38. Zhang, J., Coulston, R. J., Jones, S. T., Geng, J., Scherman, O. A. and Abell, C. (2012). One-step fabrication of supramolecular microcapsules from microfluidic droplets, *Science*, **335**, pp. 690–694.
39. Sigle, D., Kasera, S., Herrmann, L. O., Palma, A., de Nijs, B., Benz, F., Mahajan, S., Baumberg, J. J. and Scherman, O. A. (2016). Observing single molecules complexing with cucurbit[7]uril through nanogap surface-enhanced Raman spectroscopy, *J. Phys. Chem. Lett.*, **7**, pp. 704–710.
40. Peveler, W. J., Jia, H., Jeen, T., Rees, K., Macdonald, T. J., Xia, Z., Chio, W.-I. K., Moorthy, S., Parkin, I. P., Carmalt, C. J., Algar, W. R. and Lee, T.-C. (2019). Cucurbituril-mediated quantum dot aggregates formed by aqueous self-assembly for sensing applications, *Chem. Commun.*, **55**, pp. 5495–5498.

Chapter 4

Optical Spectroscopy

This chapter is designed to provide in-depth information on optical spectroscopies based on absorption and scattering of light including ultraviolet-visible (UV-Vis) spectroscopy, infrared (IR) spectroscopy, dynamic light scattering (DLS) technique, Raman spectroscopy, fluorescence spectroscopy, surface-enhanced Raman spectroscopy, and surface-enhanced fluorescence which are particularly useful for biological and chemical sensing.

4.1 Introduction

Optical spectroscopy is the study of interaction between matter and electromagnetic (EM) radiations at different frequencies, ranging from UV to IR (Fig. 4.1). The light-matter interaction can be described by the Plank-Einstein relation,

$$\Delta E = hv \tag{4.1}$$

where E is energy, h is Planck's constant, and v is the frequency of the EM radiation.

As EM radiations have characteristic energies corresponding to different physical phenomena, they are used to study the properties and structures of the system. For instance, UV-Vis photons have

Macrocycle-Functionalised Nanosensors
Weng-I Katherine Chio and Tung-Chun Lee

ISBN 978-981-5129-33-5 (Hardcover), 978-1-003-51057-4 (eBook)
www.jennystanford.com

energy in the range of electronic transitions which allows us to probe the electronic band structure or energy levels of molecular orbitals. The energy of the IR photons is in the range of bond vibrations which can be used to deduce the molecular structure of the analyte molecules (Fig. 4.1). In the context of sensing, the unique spectral signals from the molecules enable us to identify and even quantify particular targets in the samples of interest.

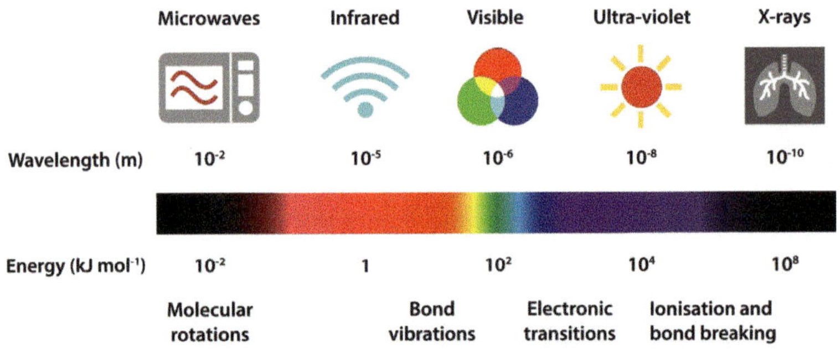

Figure 4.1 Schematic illustration of the electromagnetic spectrum, showing the energies and material phenomena across a range of wavelengths (not to scale).

The following sections will discuss each of the selected spectroscopic methodologies in detail.

4.2 UV-Visible Spectroscopy

Among all optical spectroscopic techniques, UV-Vis spectroscopy is one of the simplest methods used to detect and identify organic compounds such as biomarkers and drug molecules in the pharmaceuticals and beverage industry. UV-Vis spectroscopy is a light absorption technique which measures the amount of light absorbed by the analyte molecule over a range of wavelengths. Sensors based on UV-Vis spectroscopy are limited to molecules containing chromophores that absorb light in the UV-Vis region. When UV or visible light is absorbed, the low energy electrons in chromophores are excited to a higher energy state where energy difference is dependent on the molecular structure [1]. The absorbance bands at specific wavelengths therefore result from electronic transitions of the light absorbing molecules.

The absorbance, A, is related to the concentration of the analyte molecules present in the sample according to the Beer-Lambert law,

$$A = log_{10}\left(\frac{I_0}{I}\right) = \varepsilon \cdot c \cdot d \tag{4.2}$$

where I_0 is intensity of incident light, I is intensity of transmitted light, ε is extinction coefficient, c is concentration of light absorbing molecules, and d is the optical path length. The extinction coefficient is a material-dependent quantity which is also dependent on the wavelength [1].

A UV-Vis spectrophotometer consists of a light source (~190–900 nm), a monochromator, sample and reference cells, sets of mirrors and slits, and a detector (Fig. 4.2). A xenon lamp can be used as a single light source for both UV and visible light. Alternatively, a deuterium lamp and a tungsten-halogen lamp can be used for UV and visible light, respectively. A monochromator based on diffraction gratings can be used to select the wavelength of the incident light. The incident light passes through the sample and the transmitted light is measured by the detector. For liquid samples, they are measured in cuvettes made from plastic, glass, or quartz depending on the wavelength of incident light used. A blank solution without the analyte molecules is used as a reference. While UV-Vis spectroscopy is a versatile technique for a wide range of applications, it has low sensitivity and selectivity.

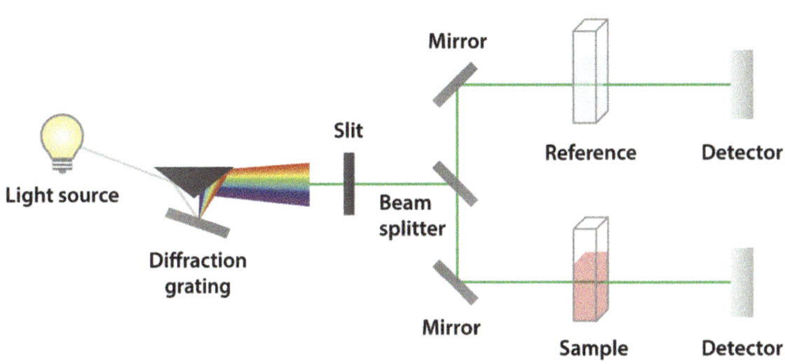

Figure 4.2 Schematic illustration of an UV-Vis spectrophotometer.

4.3 Fluorescence Spectroscopy

Fluorescence spectroscopy is based on the emission of light from fluorophores, fluorescent molecules that contain chromophores. A fluorometer (a.k.a. fluorescence spectrometer) is used to measure the intensity of fluorescence emission as a function of wavelengths. Similar to the UV-Vis spectrophotometer, a xenon lamp is used as the light source to illuminate the sample after passing through the diffraction grating (excitation monochromator). The light emitted by the analyte molecules then passes through a second diffraction grating (emission monochromator) located at 90 degrees to the first diffraction grating and measured by a detector (Fig. 4.3) [2].

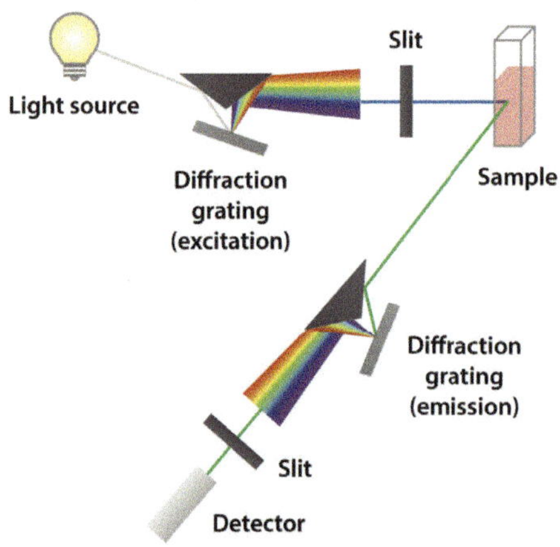

Figure 4.3 Schematic illustration of a fluorometer.

Fluorescence is a type of luminescence that occurs when the incident light excites a fluorophore from the ground state to an excited electronic state. The molecule quickly relaxes to the lowest vibrational level of an excited state within picoseconds (10^{-12} s) and loses thermal energy through vibrations. This is known as internal conversion. When it falls from the excited singlet state to the ground state, a photon is released with a lower energy but a longer wavelength than the incident photon (Fig. 4.4). The difference in energy is known as the Stokes shift. Fluorescence has a very short lifetime of around 10 ns as the return of the excited electron to the ground state is allowed by the opposite spin of electrons in the excited singlet state and the ground state [3]. Phosphorescence occurs if the photon is emitted from an excited triplet state where the spin of the excited electron is the same as the one in the ground state. The transition of an excited electron from the triplet state to the ground state is forbidden, so the lifetime of phosphorescence is much longer than fluorescence (greater than microseconds). As the triplet state has a lower energy than the singlet state, the photons emitted in phosphorescence have longer wavelengths than those in fluorescence. The electrons can also go through conversion from singlet to triplet state, which is known as intersystem crossing.

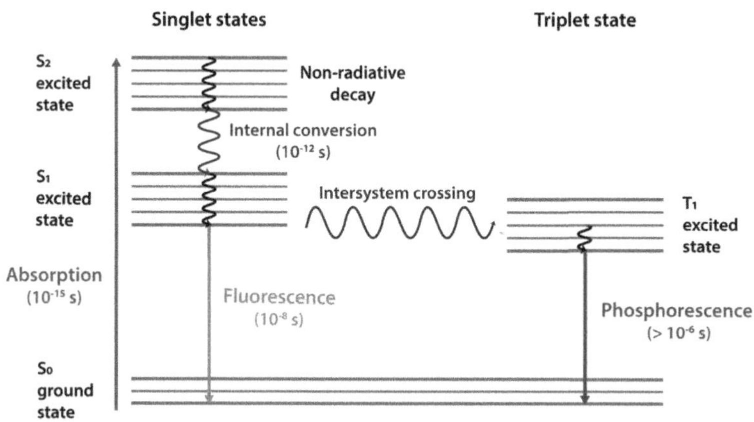

Figure 4.4 The Jablonski diagram showing fluorescence and phosphorescence.

Compared to UV-Vis spectroscopy, fluorescence spectroscopy is more sensitive and specific, and hence it is used for the detection of biomolecules and chemicals such as amino acids, proteins, organic dyes, and pigments in clinical and environmental applications [4]. Fluorescence can be affected by a number of factors such as temperature, pH, and molecular interactions (energy transfer and quenching).

4.4 IR Spectroscopy

IR spectroscopy is an absorption technique that is commonly used in forensic science and homeland security to identify compounds such as fibres, drugs, and explosives. Unlike UV-Vis spectroscopy,

IR measures the small energy absorbance between the vibrational and rotational states instead of the larger energy absorbances from electronic transitions [5]. A molecule needs to have a change in electric dipole moment during bond vibrations and rotations in order to be IR-active, which is known as the selection rule for IR spectroscopy. For instance, centrosymmetric molecules such as O_2 are IR-inactive. Molecular vibrations can be classified into stretching (changing bond length) and bending (changing bond angle). The former can be further divided into symmetric and asymmetric stretching while the latter includes scissoring, rocking, wagging, and twisting. These molecular vibrations absorb the incident IR radiation and result in a spectrum showing characteristic properties for different molecular species. The IR spectrum shows the % transmittance against wavenumber, $\tilde{\nu}$, the inverse of the wavelength, λ,

$$\tilde{\nu} = \frac{1}{\lambda} \tag{4.3}$$

which is usually given in cm^{-1} for spectroscopy.

There are two regions, namely, functional group frequency region and fingerprint region, in the spectrum. The functional group frequency region (4000–1500 cm^{-1}) can be used to deduce the functional groups present in the molecules while the fingerprint region (1500–400 cm^{-1}) can be used to characterise the analyte molecules by comparing them with known samples.

Dispersive IR and Fourier transform IR (FTIR) are the two types of IR spectrometer, which operate in the frequency and time domains, respectively. An IR spectrometer generally includes an IR radiation source, set of slits or mirrors, sample holder, detector, and signal processor. Near- (14,000–4000 cm^{-1}), mid- (4000–400 cm^{-1}), or far-IR (400–30 cm^{-1}) can be used as the radiation source, depending on the analyte molecules of interest. For example, a heated silicon carbide element is used as the IR source [6]. An interferometer, which has a fixed mirror and a movable mirror perpendicular to each other and a beam splitter, is used in FTIR to split a beam of radiation into two beams and produce an optical path between them (Fig. 4.5). The two beams then recombine and interfere with the beam splitter [6]. The intensity of radiation as a function of optical path difference is shown on an interferogram, which can be converted into a spectrum using the Fourier transform. IR spectroscopy can be used for solid and liquid samples, but pre-treatment before measurements is often challenging. Liquid samples are either sandwiched between two NaCl plates or used with AgCl or AgBr disposable windows [5]. Solid samples can be prepared by grinding and dispersing into a solid or liquid matrix to form a mull, or by grinding with dry KBr to form a thin disk. It is important to choose solvents that do not absorb and obscure spectral signals in the region of interest.

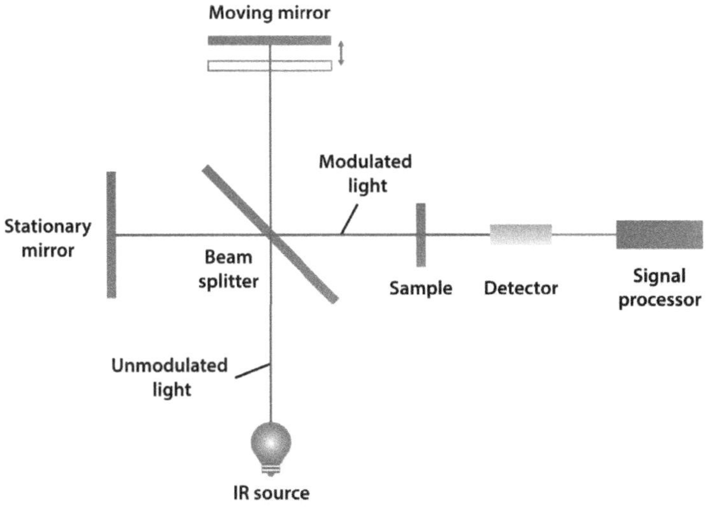

Figure 4.5 Schematic illustration of a FTIR spectrometer.

4.5 Dynamic Light Scattering Technique

Dynamic light scattering (DLS), also known as photon correlation spectroscopy or quasi-elastic light scattering, is a technique used to probe the size distribution of molecules and particles in a solution or suspension based on Brownian motion [7]. Strictly speaking, DLS is a light scattering measurement rather than a spectroscopic measurement. In particular, a monochromatic (helium-neon) laser is used as a light source and passed through a sample of colloidal solution in a plastic cuvette. The light scattered by the particles in the solution is collected by a detector at a side-scattering angle of 90 degrees or at a back-scattering angle of 173 degrees to the incident light (Fig. 4.6). A correlator is used to measure the correlation in signals over time.

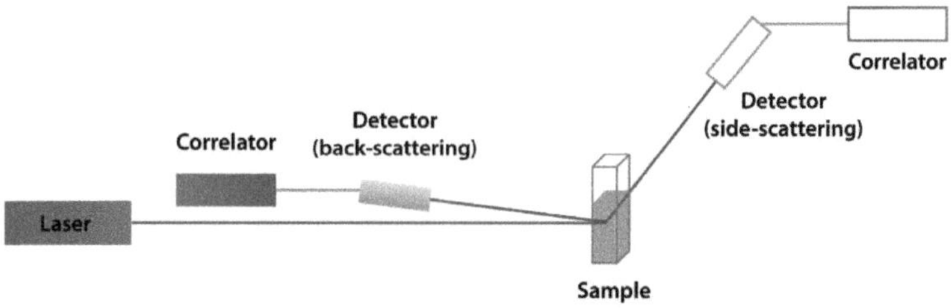

Figure 4.6 Schematic illustration of the dynamic light scattering technique.

As the Brownian motion of the particles in the solution is dependent on the size and diffusion properties of the particles, the scattered light from different particles interferes, leading to fluctuations in the light intensity. In other words, the diffusion coefficient and size distribution (polydispersity) of the particles can be obtained from the temporal fluctuations of scattered light intensity using autocorrelation and the Stokes-Einstein equation:

$$d_H = \frac{k_B T}{3\pi\eta D} \tag{4.4}$$

where d_H is hydrodynamic diameter, k_B is Boltzmann constant, T is absolute temperature, η is viscosity, and D is diffusion coefficient.

DLS data shows the scattered light weighted size distribution of particles (usually in the range from 1 nm to 10 μm) which can also be converted to volume or number distribution. The rapid, non-invasive, and versatile nature of DLS has found itself a range of applications such as characterising nanoparticles, polymers, micelles, and proteins for the pharmaceuticals and food industry.

4.6 Raman Spectroscopy

Raman spectroscopy (RS), which is a light scattering technique developed by C. V. Raman in 1928, is used to identify and differentiate materials via their vibrational fingerprints in a similar way to infrared (IR) spectroscopy [8]. In RS, a single mode laser with a wavelength ranging from UV through visible to near-IR is commonly used as the light source, which illuminates the sample after passing through sets of mirrors and slits (Figs. 4.1 and 4.7). Most photons are Rayleigh scattered while a very small number (1 in 10^6) are Raman scattered and collected by the CCD detector. The probability of Raman scattering (and hence scattering cross section) is very low as it depends on whether the incident light is in resonance with the energy transition between the ground and excited electronic states of the molecule [9].

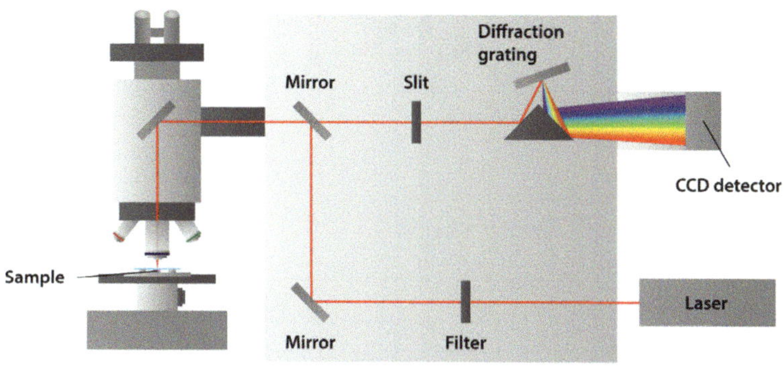

Figure 4.7 Schematic illustration of the set-up of Raman spectroscopy.

In Rayleigh scattering, the incident photon polarises the electron cloud of the analyte molecule and excites it from the ground state to the virtual state. The molecule quickly relaxes and returns to the ground state, releasing a photon of the same energy as the incident photon. This means Rayleigh scattering is an elastic process where the scattered light has the same wavelength as the incident light. In Raman scattering, the energy of the scattered photon is different from the incident photon, which means it is an inelastic process. Raman scattering can be classified into Stokes and anti-Stokes Raman scattering. When the excited molecule at the virtual state drops back to an energy level higher than the ground state, known as the vibrational state, the scattered photon has lower energy than the incident photon and thus a longer wavelength. This is called Stokes Raman scattering. When the excited molecule falls to the ground state from one of the virtual states, the scattered photon has a higher energy than the incident photon and a shorter wavelength. This is known as anti-Stokes Raman scattering (Fig. 4.8). The intensity of the anti-Stokes Raman scattering is lower than the Stokes Raman scattering, which can be explained by the Boltzmann distribution,

$$p \propto \exp\left(\frac{-E}{k_B T}\right) \tag{4.5}$$

where p is the probability of finding a molecule in a particular energy state, E is energy, k_B is Boltzmann constant, and T is temperature.

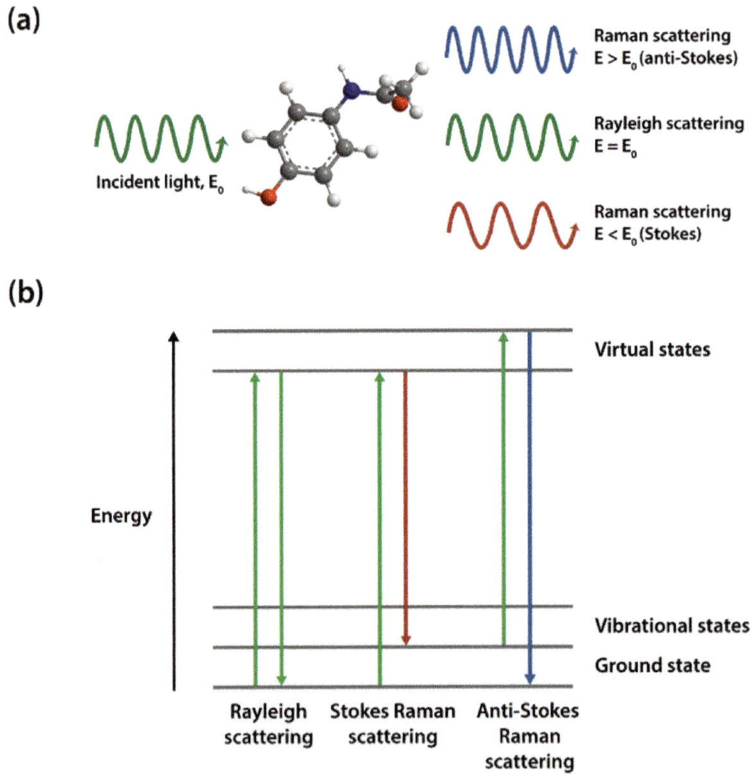

Figure 4.8 Schematic illustrations of (a) Rayleigh and Raman scattering by a molecule (paracetamol) and (b) the corresponding energy transitions.

As there is a higher probability of finding a molecule in the ground state than in the vibrational state, the Stokes Raman scattering is more intense than the anti-Stokes Raman scattering and hence more commonly used in RS [10]. The Raman spectrum shows the intensity of photons collected as a function of the Raman shift, i.e., the energy difference between the incident and Raman scattered photons, which is given in wavenumber, cm^{-1}. However, only the Raman-active modes of the analyte molecule appear in the spectrum. The selection rule for RS is that molecular vibration needs to be coupled to a change in the polarisability of the molecule instead of a change in the dipole moment as in IR spectroscopy [11]. Polarisability measures the ease of distortion of the electron cloud of the molecule, which is dependent on the bond length, with the shorter bonds being more difficult to polarise. For molecules with a centre of symmetry, the Raman-active modes are IR-inactive and vice versa. This is called the rule of mutual exclusion and hence the two techniques are complementary.

RS offers numerous advantages such as non-destructive, no sample preparation, fast response, high spatial resolution, and tolerance to water and CO_2 while providing very detailed material information, thus meriting its applications in different areas including pharmaceuticals, geology, carbon materials, semiconductors, and polymer analysis. However, RS is not suitable for diluted samples due to its inherently weak signals, thus limiting its application to bulk materials. The Raman signal can also be hindered by fluorescence from the background matrix. Several techniques based on RS were subsequently developed to tackle this problem.

4.7 Surface-Enhanced Raman Spectroscopy

Surface-enhanced Raman scattering (SERS) was first observed by Fleischmann and co-workers when they were measuring the Raman scattering of pyridine on a roughened Ag electrode in 1974. The SERS phenomenon was then interpreted by two groups, Jeanmaire and Van Duyne and Albrecht and Creighton, in 1977. Over the years, SERS has sparked great interest across different disciplines including chemistry, biology, materials science, and nanotechnology. SERS is a very powerful sensing technique that can be used to quantify trace analyte molecules in a wide range of applications such as disease diagnostics, therapeutic drug monitoring, environmental monitoring, and homeland security (see Chapters 5–7).

In SERS, nanoparticles (NPs) or nanostructured surfaces such as Au and Ag are used as substrates to overcome the issue of intrinsically low scattering cross sections in Raman spectroscopy, thus resulting in greatly enhanced Raman signals. An enhancement factor (EF) of ~10^5 to 10^6 can be reached on average while single-molecule detection is possible in some cases. The overall EF can be attributed to an EM enhancement (see Section 4.7.2) of the incident light absorbed on or near the surface of NPs via localised surface plasmon resonance (LSPR, see Section 4.7.1), and a chemical (CHEM) enhancement (see Section 4.7.3) associated with an increase in polarisability of the analyte molecules during charge transfer between the substrate and the analyte molecules (Fig. 4.9).

The overall EF (G_{SERS}) is given by [12]:

$$G_{\text{SERS}} = G_{\text{EM}} G_{\text{Chem}} \tag{4.6}$$

where G_{EM} and G_{Chem} are the contributions from the EM enhancement and CHEM enhancement, respectively.

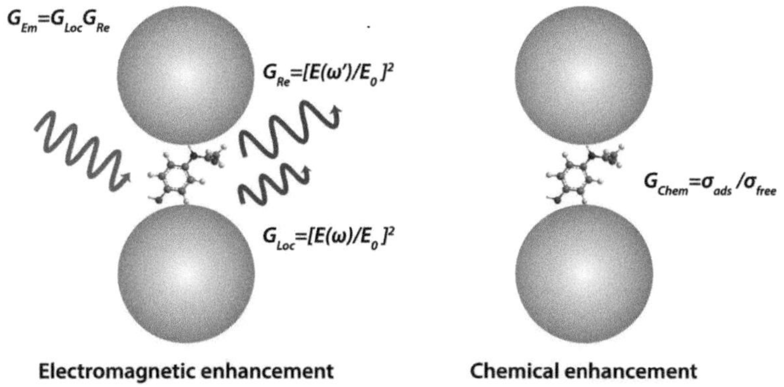

Figure 4.9 Schematic illustration of the electromagnetic enhancement (local electric field enhancement and dipole re-radiation enhancement) and the chemical enhancement in SERS [13].

4.7.1 Surface Plasmon Resonance

Surface plasmon resonance (SPR) is a photophysical process in which the free (conduction band) electrons within the NPs are polarised by the electric field of the incident light to create a coherent multipolar (e.g., dipolar) oscillation between the opposite charges of the NPs (Fig. 4.10), giving the plasmonic NPs their distinct colour [14]. When the coherent oscillation is highly localised to a particular position (e.g., on the NP surface), it is known as localised surface plasmon resonance (LSPR).

In 1908, Mie was the first to explain the ruby-red colour of Au NPs by providing an analytical solution to Maxwell's equations in the non-conducting medium for the interaction between an incident EM field and a homogeneous spherical NP [15]. Maxwell's equations are as follows:

$$\nabla \cdot \boldsymbol{E} = 0 \tag{4.7}$$

$$\nabla \cdot \boldsymbol{H} = 0 \tag{4.8}$$

$$\nabla \times \boldsymbol{E} = i\omega\mu\boldsymbol{H} \tag{4.9}$$

$$\nabla \times \boldsymbol{H} = -i\omega\varepsilon\mathbf{E} \tag{4.10}$$

where $\boldsymbol{E}$ and $\boldsymbol{H}$ are the complex electric field and magnetic field of the incident electromagnetic field, ω is angular frequency of the EM field, ε is the dielectric constant and μ is magnetic permeability of the medium [16].

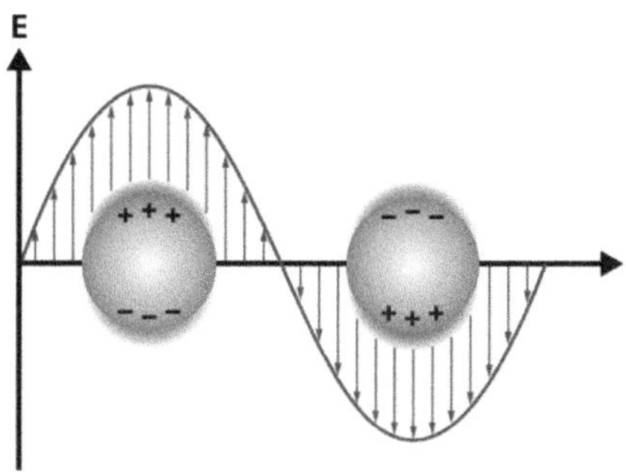

Figure 4.10 Schematic illustration of localised surface plasmon resonance of Au NPs.

They can be expressed as two Helmholtz equations,

$$\nabla^2 \boldsymbol{E} + k^2 \boldsymbol{E} = 0 \tag{4.11}$$

$$\nabla^2 \boldsymbol{H} + k^2 \boldsymbol{H} = 0 \tag{4.12}$$

where k is a wavenumber given by

$$k^2 \equiv \omega^2 \varepsilon\mu \tag{4.13}$$

The extinction cross section σ_{ext} of a spherical NP, which is the sum of its absorption cross section σ_{abs} and scattering cross section σ_{sca}, is then obtained by solving the equations above [13, 16]:

$$\sigma_{ext}(\omega) = \sigma_{abs}(\omega) + \sigma_{sca}(\omega) \tag{4.14}$$

$$\sigma_{abs}(\omega) = 4\pi k r^3 i\left[\frac{\varepsilon(\omega) - \varepsilon_m}{\varepsilon(\omega) + 2\varepsilon_m}\right] \tag{4.15}$$

$$\sigma_{sca}(\omega) = \frac{8\pi}{3} k^4 r^6 \left|\frac{\varepsilon(\omega) - \varepsilon_m}{\varepsilon(\omega) + 2\varepsilon_m}\right|^2 \tag{4.16}$$

where r is radius of the NP, $\varepsilon(\omega)$ is complex dielectric constant of the NP and ε_m is dielectric constant of the surrounding medium.

If the radius of the NP is much smaller than the wavelength of the incident light (r < 10 nm), the NP experiences a homogeneous electric field (time-dependent but spatial-independent) under the quasi-state approximation. The electric field inside the NP (E_{in}) is then given by [13,16]:

$$E_{in} = E_0 \frac{3\varepsilon_m}{\varepsilon(\omega) + 2\varepsilon_m} \tag{4.17}$$

where E_0 is the incident electric field.

In this case, a dipolar charge distribution is generated on the surface of the NP. The extinction cross section is mainly contributed by the absorption cross section (Fig. 4.11a) which can then be described by [16, 17]

$$\sigma_{ext}(\omega) = \frac{9\varepsilon_m^{3/2} V\omega}{c} \frac{\varepsilon_2(\omega)}{\left(\varepsilon_1(\omega) + 2\varepsilon_m\right)^2 + \varepsilon_2^2(\omega)} \tag{4.18}$$

where V is the volume of the NP, c is the speed of light, ε_1 and ε_2 are the real and imaginary parts of the dielectric constant of the NP.

If ε_2 is sufficiently small, σ_{ext} is maximised at a resonance frequency such that

$$\varepsilon_1(\omega) = -2\varepsilon_m \tag{4.19}$$

The position and bandwidth of the LSPR peaks are determined by real and imaginary parts of the dielectric constant of the NP [17]. For instance, the LSPR peaks of Au NPs and Ag NPs lie within the visible range of the EM spectrum, at ~520 and ~420 nm, respectively. The optical losses of the NPs, which are mainly due to interband and intraband transitions, are also described by the imaginary part of the dielectric constant [18]. The localised surface plasmons of Au NPs can decay into electron-hole pairs due to overlapping of LSPR frequencies and band edge, and they have higher optical losses than Ag NPs in the visible range [19]. In contrast, the frequency at which interband transition becomes allowed in Ag NPs is above the LSPR frequency. The localised surface plasmons cannot decay into the electron-hole pairs, thus resulting in lower losses and stronger SERS signals [19].

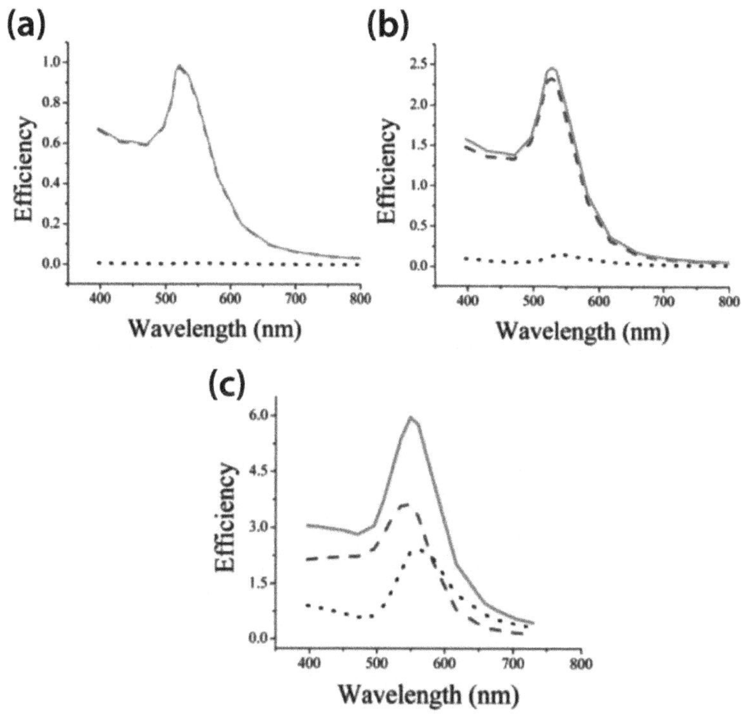

Figure 4.11 Efficiency of absorption (red dashed), scattering (black dotted), and extinction (green solid) for Au NPs with a diameter of (a) 20 nm, (b) 40 nm, and (c) 80 nm [20].

If the size of the NPs is greater than their electron mean free path (e.g., ~40 nm for Au), the contribution from scattering starts to appear and the extinction cross section becomes more complex (Fig. 4.11b) [17, 20]. When the size of the Au NPs reaches ~80 nm, the contributions of absorption and scattering to the extinction cross section become comparable (Fig. 4.11c). As the particle size increases, the extinction cross section increases, resulting in red-shift and broadening of the LSPR peak (Fig. 4.12a).

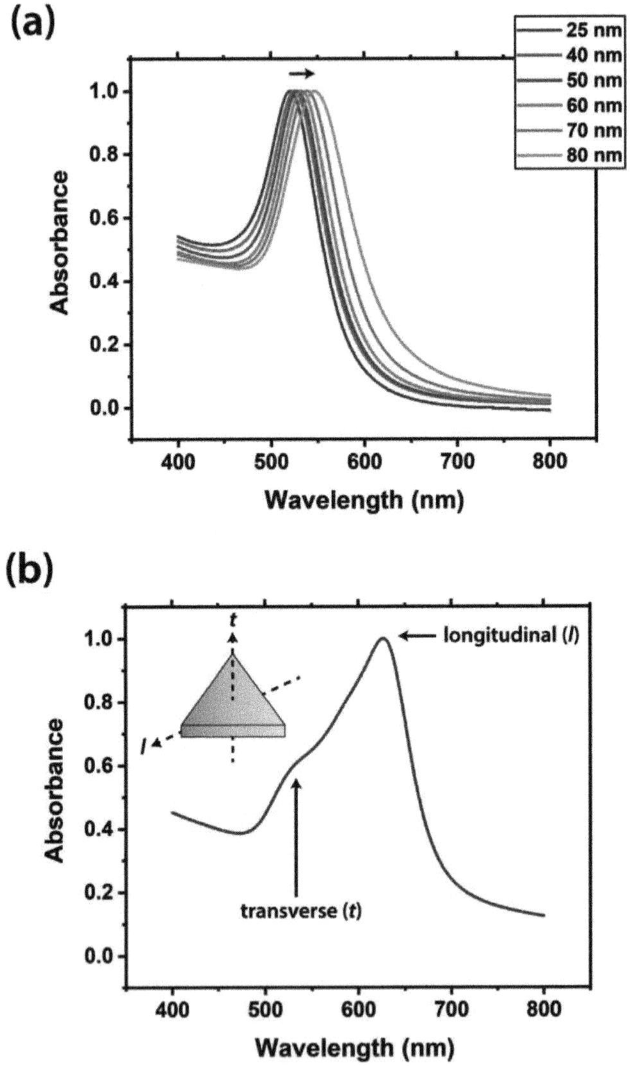

Figure 4.12 (a) UV-Vis spectra of Au NPs with sizes from 25 to 80 nm, showing gradual red-shifts in the LSPR peak as the particle size increases. (b) UV-Vis spectra of Au NTs showing the transverse and longitudinal LSPR peaks.

When the shape of the NPs derivates from sphericity, the LSPR in the transverse and longitudinal directions are no longer equivalent. For instance, there are two distinct LSPR peaks in the UV-Vis spectrum of Au nanotriangles (Au NTs), with the lower and higher wavelengths corresponding to the transverse and longitudinal LSPR, respectively (Fig. 4.12b) [16]. Herein, the LSPR and hence EM enhancement of SERS are dependent on the dielectric constant, size, and shape of the NPs. For biomedical sensing, the penetrating depth can be improved by tuning the wavelength of the LSPR peak into the NIR range (600–900 nm), which is advantageous over fluorescence spectroscopy as no UV excitation is required [21].

4.7.2 EM Enhancement

The EF in SERS is mainly contributed by the EM enhancement ($\sim 10^4$–10^{11}), which is a two-step process involving local electric field and re-irradiation enhancements resulting from the LSPR of the NPs with the incident EM field [12, 13, 22]. When the analyte molecules are located in close proximity to the NPs, the local electric field (E) oscillating at an incident frequency of ω is enhanced. The local electric field enhancement (G_{Loc}) is defined by:

$$G_{Loc} = \left[\frac{E(\omega)}{E_0}\right]^2 \quad (4.20)$$

where E_0 is the incident electric field.

The NPs then transfer the enhanced Raman signals, which are proportional to the square of the enhanced local electric field at a scattered frequency of ω', from the near field to the far field. This is due to the mutual excitation between the dipole of the NPs and the induced dipole of the molecules. The dipole re-irradiation enhancement (G_{Re}) is defined by:

$$G_{Re} = \left[\frac{E(\omega')}{E_0}\right]^2 \quad (4.21)$$

For small Raman shifts, i.e., a small difference in frequency between incident and Raman-scattered photons, the EM enhancement is approximately proportional to the fourth power of the local electric field enhancement.

$$G_{Em} = G_{Loc} G_{Re} = \left[\frac{E(\omega)}{E_0}\right]^2 \left[\frac{E(\omega')}{E_0}\right]^2 = \left[\frac{E(\omega)}{E_0}\right]^4 \quad (4.22)$$

This $|E(\omega)|^4$ approximation is commonly used and valid for most applications. However, one may need to account for the non-local effects caused by the sub-nm spacing between the NPs (quantum confinement effects), the anisotropic nanostructures, and the random orientations of molecules (polarisation effects) under certain circumstances [13, 23–25].

In addition, the EM enhancement is strongly dependent on the distance from the surface of the NP (d). The mathematical relationship is given by

$$\frac{G_{Em}(d)}{G_{Em}(0)} = \left[\frac{r}{r+d}\right]^{12} \tag{4.23}$$

where r is the radius of the NP.

The SERS enhancement decreases significantly for a small increase in distance (Fig. 4.13). Thus, it is very important to bring the analyte molecules close to the surface of the SERS substrate for sensing applications.

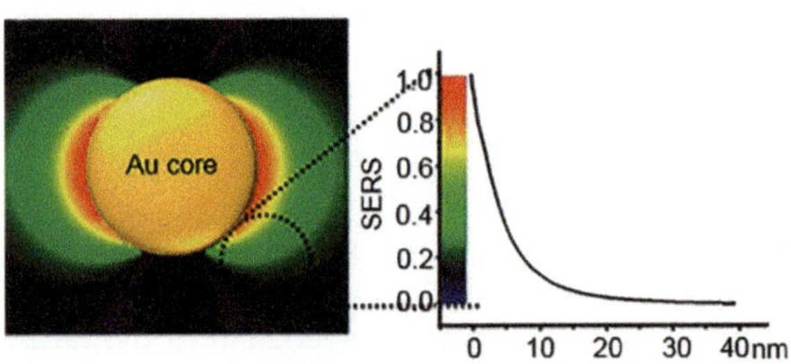

Figure 4.13 Finite difference time domain simulation of spatial distribution of local electric field of an Au NP [22].

4.7.3 CHEM Enhancement

CHEM enhancement has a much smaller contribution to the EF (10–100) compared to the EM enhancement. It is still important to understand the chemical mechanisms in order to optimise the SERS signals for sensing applications [9, 22]. When the analyte molecules are adsorbed onto the NPs, the polarisability and the cross sections of the vibration modes are modified via surface interaction with the substrate [9–12]. The CHEM enhancement (G_{Chem}) is defined by

$$G_{\text{Chem}} = \frac{\sigma_{\text{ads}}}{\sigma_{\text{free}}} \quad (4.24)$$

where σ_{ads} and σ_{free} are the cross sections of adsorbed and free (unbound) molecules, respectively.

There are three mechanisms contributing to the overall CHEM enhancement, the non-resonant interaction (ground state interaction between molecule and NP), the molecular resonance (resonance of incident light with molecule), and the charge transfer resonance (resonance of incident light with molecule-metal coupling) [26, 27].

The CHEM enhancement can be calculated using electronic structure methods such as time-dependent density functional theory (DFT). The static polarisability derivatives of the molecule-metal system (ground state) are calculated by modelling the NPs as clusters of atoms. However, this method fails to interpret the weakly coupled (long-range) charge transfer correctly as it is strongly mixed with LSPR [27, 28]. Semi-empirical molecular orbital methods such as intermediate neglect of differential overlap Hamiltonian and the single configuration interaction approaches are then applied to describe the CHEM enhancement from charge transfer by separating the LSPR and charge transfer resonance in the calculations [9, 29].

4.7.4 Hotspots

The SERS EF is dependent on the geometry of the nanostructured substrate, or inter-particle spacing between the NPs, and the distance and relative orientation between analyte molecules and the substrate. The EM enhancement is highly localised in a very small spatial volume known as a hotspot. Plasmonic hotspots are located at the nanogaps between two NPs, sandwiched between a NP and a thin film, or at the sharp corners or tips of nanostructures. The local electric field enhancement is much stronger in the nanogaps than on the surface of NPs. Therefore, aggregated NPs with well-defined plasmonic hotspots are preferred over isolated NPs for SERS substrates.

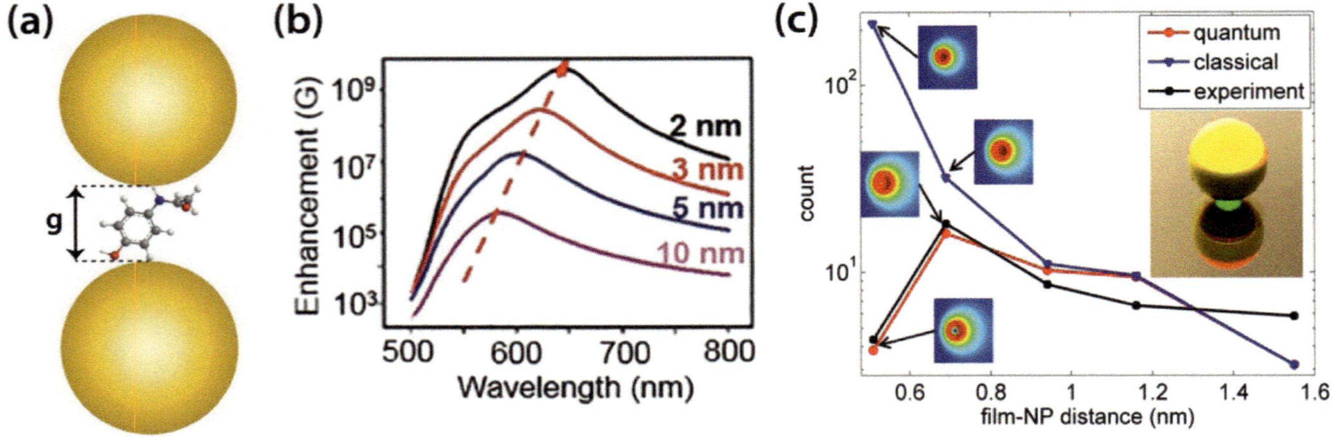

Figure 4.14 (a) Schematic illustration of a nanogap between Au NPs. (b) The dependence of EM enhancement on the size of the nanogap between Au NPs. (c) The quantum tunnelling effect on the EM enhancement between Au NP and Au film [22, 30].

When the analyte molecules are located in close proximity to the plasmonic hotspots, the SERS signals are enhanced significantly, thus meriting SERS sensing applications. For instance, the SERS EF increases by four orders of magnitudes when the nanogap between the Au NP dimers is reduced from 10 to 2 nm (Fig. 4.14a,b) [13, 31, 32]. The dependence of EM enhancement on the size of nanogap (g) is given by a power-law [32]

$$G_{\text{EM}} \sim \frac{1}{g^n} \tag{4.25}$$

where $n \sim 2$.

When the nanogap is at a sub-nm scale (<0.7 nm) between Au NPs and Au film, the EF can be reduced by an order of magnitude due to a non-local effect known as quantum tunnelling (Fig. 4.14c) [30, 32, 33].

4.8 SERS-Related Techniques

In the past years, a number of optical techniques based on or related to SERS have been developed, including, but not limited to, resonance Raman spectroscopy (RRS), coherent anti-Stokes Raman spectroscopy (CARS), surface-enhanced hyper Raman spectroscopy (SEHRS), surface-enhanced Raman optical activity (SEROA), surface-enhanced spatially offset Raman spectroscopy (SESORS), electrochemical surface-enhanced Raman spectroscopy (EC-SERS), tip-enhanced Raman spectroscopy (TERS), shell-isolated nanoparticle-enhanced Raman spectroscopy (SHINERS), and surface-enhanced fluorescence spectroscopy (SEFS).

RRS is an analytical technique based on RS in which the frequency of the incident photons is close to or overlaps with that of an electronic transition of the analyte molecules. The electrons are excited to higher electronic states and a subset of Raman-active modes is greatly enhanced due to resonance effects [34]. This technique can be used for selective monitoring of molecules in a complex matrix, for instance, biological chromophores embedded in membrane proteins [35]. CARS is a non-linear optical technique that utilises two lasers at different frequencies to generate vibrational coherences where the frequency of a phonon (vibrational mode) is equal to the frequency difference between two photons [35]. Due to its intrinsic vibrational contrast, CARS is used for biomedical imaging and remote sensing as it does not require staining or chemical labelling [36]. SEHRS, which is a non-linear optical analogue of SERS excited by two photons, provides complementary spectral information owing to different selection rules [37]. It may be used to probe Raman-inactive modes and even those which are both Raman-inactive and IR-inactive [37]. The applications of SEHRS include biosensing and pH sensing.

SEROA is a chiral-sensitive vibrational technique integrating SERS and Raman optical activity (ROA), where a tiny difference in the Raman intensity of a chiral molecule in left- and right-circularly polarised light is measured [38]. This technique can be potentially used in protein and RNA studies [39]. SESORS combines SERS and spatially offset Raman spectroscopy (SORS) which is a non-invasive method used to probe subsurface information by applying a spatial offset between excitation and collection points to collect the photons scattered by the subsurface [40]. SESORS is particularly useful for biosensing as it allows the detection of SERS-labelled analyte molecules in deep tissues with improved sensitivity.

EC-SERS utilises electrochemical potential to enhance the SERS signals of analyte molecules. When the energy of the incident photons overlaps with the transition between Fermi levels in molecule-metal couplings, charge transfer occurs, resulting in CHEM enhancement [9]. This technique has been used for the detection of pathogens and electrocatalytic reactions [41, 42]. TERS, which couples SERS with scanning probe microscopy, utilises a plasmonic tip at the nanoscale to generate a strong EM enhancement at the tip apex. This increases the spatial resolution to a few nm with intramolecular chemical resolution for biological samples such as bacteria, viruses, and DNA/RNA crystals [9, 43].

SHINERS is a relatively new technique based on SERS that uses smart dust such as Au@SiO_2 or Au@Al_2O_3 core-shell NPs as substrates [44]. The chemically inert shell of the plasmonic NPs can protect them from aggregating or direct contact with the analyte molecules. This technique allows us to probe various types of molecules including pesticide residues, food additives, and cancer cells anchored on different objects with strong SERS enhancement [45]. SEFS, also known as plasmon-enhanced fluorescence (PEF), is a successor of fluorescence spectroscopy (see Section 4.3), offering advantages such as enhanced fluorophore emission and photostability [46]. SEFS makes use of the significant EM enhancement from the surface of plasmonic NPs to increase the extinction cross section by a few orders of magnitude and decrease fluorescence lifetime. Nanosensors based on SEFS have been used to detect DNA, adenosine-5'-triphosphate (ATP), and heavy metal ions [47].

References

1. Perkampus, H.-H. (1992). *UV-Vis Spectroscopy and Its Applications* (Springer-Verlag).
2. Gomes, A. J., Lunardi, C. N., Rocha, F. S. and Patience, G. S. (2019). Experimental methods in chemical engineering, fluorescence emission spectroscopy, *Can. J. Chem. Eng.*, **97**, pp. 2168–2175.
3. Lichtman, J. W. and Conchello, J.-A. (2005) Fluorescence microscopy, *Nat. Methods*, **2**, pp. 910–919.
4. Lakowicz, J. R. (2006). *Principles of Fluorescence Spectroscopy* (Springer-Verlag).
5. Tyner, T. and Francis, J. (2017). *Infrared Spectroscopy*, Part 2, Analytical Procedures and General Directions (ACS Reagent Chemicals).
6. Stuart, B. (2020) *Analytical Techniques in Forensic Science*, Chapter 7 "Infrared Spectroscopy" (Wiley), pp. 145–160.
7. Babick, F. (2020). *Characterization of Nanoparticles*, Chapter 3 "Dynamic Light Scattering (DLS)" (Elsevier), pp. 137–172.
8. Raman, C. V. and Krishnan, K. S. (1928). A new type of secondary radiation, *Nature*, **121**, pp. 501–502.
9. Langer, J., Jimenez de Aberasturi, D., Aizpurua, J., Alvarez-Puebla, R. A., Auguie, B., Baumberg, J. J., Bazan, G. C., Bell, S. E. J., Boisen, A., Brolo, A. G. et al. (2020). Present and future of surface-enhanced Raman scattering, *ACS Nano*, **14**, pp. 28–117.
10. Kauffmann, T. H., Kokanyan, N. and Fontana, M. D. (2019). Use of Stokes and anti-Stokes Raman scattering for new applications, *J. Raman Spectrosc.*, **50**, pp. 418–424.
11. Hallmark, V. M. and Campion, A. (1986). Selection rules for surface Raman spectroscopy: experimental results, *J. Chem. Phys.*, **84**, pp. 2933–2941.

12. Pilot, R., Signorini, R., Durante, C., Orian, L., Bhamidipati, M. and Fabris, L. (2019). A review on surface-enhanced Raman scattering, *Biosensors*, **9**, 57.
13. Ding, S.-Y., Yi, J., Li, J.-F., Ren, B., Wu, D.-Y., Panneerselvam, R. and Tian, Z.-Q. (2016). Nanostructure-based plasmon-enhanced Raman spectroscopy for surface analysis of materials, *Nat. Rev. Mater.*, **1**, 16021.
14. Cao, G. (2010) Nanostructures and Nanomaterials – Synthesis, Properties and Applications, 2nd ed. (World Scientific Publishing Company).
15. Mie, G. (1908). Beitrage zur optik truber medien, speziell kolloidaler metallosungen, *Ann. Phys.*, **330**, pp. 337–445 (in German).
16. Ghosh, S. K. and Pal, T. (2007). Interparticle coupling effect on the surface plasmon resonance of gold nanoparticles: from theory to applications, *Chem. Rev.*, **107**, pp. 4797–4862.
17. Huang, X. and El-Sayed, M. A. (2010). Gold nanoparticles: optical properties and implementations in cancer diagnosis and photothermal therapy. *J. Adv. Res.*, **1**, pp. 13–28.
18. Arnold, M. D. and Blaber, M. G. (2009). Optical performance and metallic absorption in nanoplasmonic systems, *Opt. Express*, **17**, pp. 3835–3847.
19. Amendola, V., Pilot, R., Frasconi, M., Marago, O. M. and Iati, M. A. (2017). Surface plasmon resonance in gold nanoparticles: a review, *J. Phys.: Condens. Matter*, **29**, 203002.
20. Jain, P. K., Lee, K. S., El-Sayed, I. H. and El-Sayed, M. A. (2006). Calculated absorption and scattering properties of gold nanoparticles of different size, shape, and composition: applications in biological imaging and biomedicine, *J. Phys. Chem. B*, **110**, pp. 7238–7248.
21. Kobayashi, H., Ogawa, M., Alford, R., Choyke, P. L. and Urano, Y. (2010). New strategies for fluorescent probe design in medical diagnostic imaging, *Chem. Rev.*, **110**, pp. 2620–2640.
22. Zong, C., Xu, M., Xu, L.-J., Wei, T., Ma, X., Zheng, X.-S., Hu, R. and Ren, B. (2018). Surface-enhanced Raman spectroscopy for bioanalysis: reliability and challenges. *Chem. Rev.*, **118**, pp. 4946–4980.
23. Le Ru, E. C., Grand, J., Felidj, N., Aubard, J., Levi, G., Hohenau, A., Krenn, J. R., Blackie, E. and Etchegoin, P. G. (2008). Experimental verification of the SERS electromagnetic model beyond the $|E|^4$ approximation: polarization effects. *J. Phys. Chem. C*, **112**, pp. 8117–8121.
24. Fazio, B., D'Andrea, C., Bonaccorso, F., Irrera, A., Calogero, G., Vasi, C., Gucciardi, P. G., Allegrini, M., Toma, A., Chiappe, D. et al. (2011). Re-

radiation enhancement in polarized surface-enhanced resonant Raman scattering of randomly oriented molecules on self-organized gold nanowires, *ACS Nano*, **5**, pp. 5945–5956.

25. De Abajo, F. J. G. (2008). Nonlocal effects in the plasmons of strongly interacting nanoparticles, dimers, and waveguides, *J. Phys. Chem. C*, **112**, pp. 17983–17987.

26. Lombardi, J. R. and Birke, R. L. (2009). A unified view of surface-enhanced Raman scattering, *Acc. Chem. Res.*, **42**, pp. 734–742.

27. Jensen, L., Aikens, C. M. and Schatz, G. C. (2008). Electronic structure methods for studying surface-enhanced Raman scattering, *Chem. Soc. Rev.*, **37**, pp. 1061–1073.

28. Morton, S. M. and Jensen, L. (2009). Understanding the molecule-surface chemical coupling in SERS, *J. Am. Chem. Soc.*, **131**, pp. 4090–4098.

29. Gieseking, R. L., Ratner, M. A. and Schatz, G. C. (2016). Semiempirical modeling of Ag nanoclusters: new parameters for optical property studies enable determination of double excitation contributions to plasmonic excitation, *J. Phys. Chem. A*, **120**, pp. 4532–4549.

30. Hajisalem, G., Nezami, M. S. and Gordon, R. (2014). Probing the quantum tunneling limit of plasmonic enhancement by third harmonic generation, *Nano Lett.*, **14**, pp. 6651–6654.

31. Le Ru, E. C. and Etchegoin, P. G. (2009) *Principles of Surface-Enhanced Raman Spectroscopy*, Chapter 4 "SERS Enhancement Factors and Related Topics" (Elsevier), pp. 185–264.

32. McMahon, J. M., Li, S., Ausman, L. K. and Schatz, G. C. (2012). Modeling the effect of small gaps in surface-enhanced Raman spectroscopy, *J. Phys. Chem. C*, **116**, pp. 1627–1637.

33. Zhu, W. and Crozier, K. B. (2014). Quantum mechanical limit to plasmonic enhancement as observed by surface-enhanced Raman scattering, *Nat. Commun.*, **5**, 5228.

34. Robert, B. (2009). Resonance Raman spectroscopy, *Photosynth. Res.*, **101**, pp. 147–155.

35. Virga, A., Ferrante, C., Batignani, G., De Fazio, D., Nunn, A. D. G., Ferrari, A. C., Cerullo, G. and Scopigno, T. (2019). Coherent anti-Stokes Raman spectroscopy of single and multi-layer graphene, *Nat. Commun.*, **10**, 3658.

36. Krishna, R., Unsworth, T. J. and Edge, R. (2016) *Reference Module in Materials Science and Materials Engineering*, Raman Spectroscopy and Microscopy (Elsevier).

37. Madzharova, F., Heiner, Z. and Kneipp, J. (2017). Surface enhanced hyper Raman scattering (SEHRS) and its applications, *Chem. Soc. Rev.*, **46**, pp. 3980–3999.

38. Novak, V., Sebestík, J. and Bour, P. (2012). Theoretical modeling of the surface-enhanced Raman optical activity, *J. Chem. Theory Comput.*, **8**, pp. 1714–1720.

39. Abdali, S. and Blanch, E. W. (2008). Surface enhanced Raman optical activity (SEROA), *Chem. Soc. Rev.*, **37**, pp. 980–992.

40. Asiala, S. M., Shand, N. C., Faulds, K. and Graham, D. (2017). Surface-enhanced, spatially offset Raman spectroscopy (SESORS) in tissue analogues, *ACS Appl. Mater. Interfaces*, **9**, pp. 25488–25494.

41. Wu, D.-Y., Li, J.-F., Ren, B. and Tian, Z.-Q. (2008). Electrochemical surface-enhanced Raman spectroscopy of nanostructures, *Chem. Soc. Rev.*, **37**, pp. 1025–1041.

42. Lee, S. H., Lee, W.-C., Koh, E. H., Ansah, I. B., Yang, J.-Y., Mun, C. W., Lee, S., Kim, D.-H., Jung, H. S. and Park, S.-G. (2022). Organometallic hotspot engineering for ultrasensitive EC-SERS detection of pathogenic bacteria-derived DNAs, *Biosens. Bioelectron.*, **210**, 114325.

43. Verma, P. (2017). Tip-enhanced Raman spectroscopy: technique and recent advances, *Chem. Rev.*, **117**, pp. 6447–6466.

44. Li, J. F., Huang, Y. F., Ding, Y., Yang, Z. L., Li, S. B., Zhou, X. S., Fan, F. R., Zhang, W., Zhou, Z. Y., Wu, D. Y., Ren, B., Wang, Z. L. and Tian, Z. Q. (2010). Shell-isolated nanoparticle-enhanced Raman spectroscopy, *Nature*, **464**, pp. 392–395.

45. Krajczewski, J. and Kudelski, A. (2019). Shell-isolated nanoparticle-enhanced Raman spectroscopy, *Front. Chem.*, **7**, 410.

46. Li, J.-F., Li, C.-Y. and Aroca, R. F. (2017). Plasmon-enhanced fluorescence spectroscopy, *Chem. Soc. Rev.*, **46**, pp. 3962–3979.

47. Li, M., Cushing, S. K. and Wu, N. (2015). Plasmon-enhanced optical sensors: A review, *Analyst*, **140**, pp. 386–406.

Chapter 5

Case Study: Nitroexplosives

This chapter is intended to provide two full case studies on macrocycle-functionalised nanosensors based on cucurbit[7]urils (CB7) and gold nanoparticles (Au NPs) or CdTe quantum dots (QDs) for the detection of nitroexplosives, 2,4-dinitrotoluene (DNT), and nitrobenzene (NB) via surface-enhanced Raman scattering (SERS) and fluorescent effects.

5.1 Introduction

Rapid detection of trace explosives remains one of the major challenges in homeland security and environmental monitoring around the world, especially in countries where there are ongoing wars. Explosives or potentially explosive compounds such as 2,4-dinitrotoluene (DNT) and 2,4,6-trinitrotoluene (TNT) are associated with bombs and landmines used in military activities or terrorist attacks while also present as contaminants in water and soils at military weapon manufacturing sites [1, 2]. DNT and TNT are toxic upon exposure, leading to adverse health effects such as decreased neurological, haematological, renal, and reproductive functions. Common sensors for rapid screening against nitroexplosives are currently based on techniques such as colorimetry, X-ray, and ion mobility spectrometry, which are not very selective. For instance, the colorimetric technique (e.g., Expray)

Macrocycle-Functionalised Nanosensors
Weng-I Katherine Chio and Tung-Chun Lee

ISBN 978-981-5129-33-5 (Hardcover), 978-1-003-51057-4 (eBook)
www.jennystanford.com

may not be able to differentiate certain types of explosives and is prone to false-positive (e.g., nitrate-containing fertilisers). Thus, it is essential to develop high performance nitroexplosive sensors in response to military activities and safety concerns.

5.2 CB7-Au NP Nanosensors for Nitroexplosives

Nanosensors based on macrocycle-functionalised plasmonic NPs allow rapid detection of analyte molecules through surface-enhanced Raman scattering (SERS) meanwhile offering other advantages including cost-effectiveness, high chemical stability, minimal sample preparation, and wide information window. Macrocycles can significantly enhance the signals of the nitroexplosives by controlling the inter-particle spacing between plasmonic NPs and localising them in close proximity to the hotspots via formation of host–guest complexes. This section will discuss a case on SERS sensing of DNT using nanoaggregates of cucurbit[7]urils (CB7) and gold nanoparticles (Au NPs), covering from host–guest complexation, through formation of plasmonic hotspots and SERS quantification, to selectivity and tolerance tests (Fig. 5.1) [3].

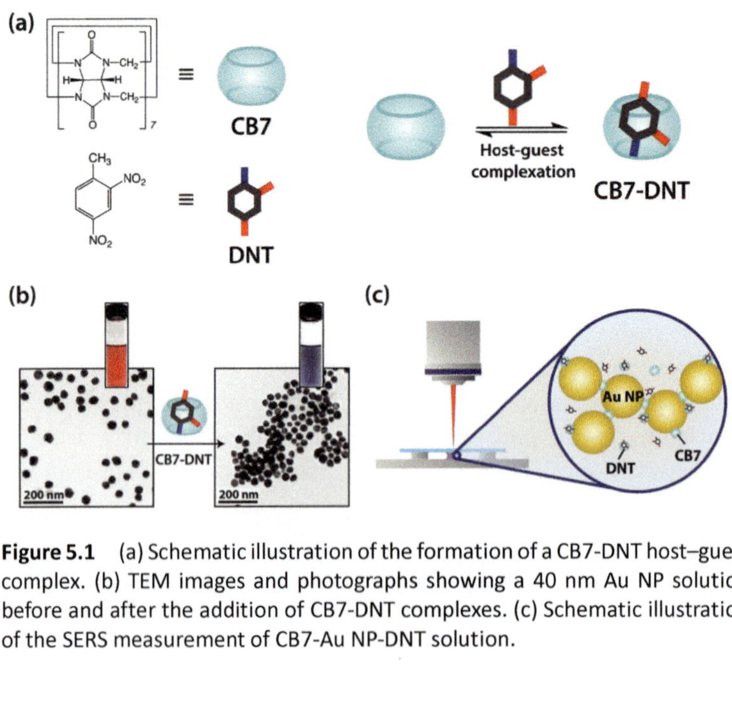

Figure 5.1 (a) Schematic illustration of the formation of a CB7-DNT host–guest complex. (b) TEM images and photographs showing a 40 nm Au NP solution before and after the addition of CB7-DNT complexes. (c) Schematic illustration of the SERS measurement of CB7-Au NP-DNT solution.

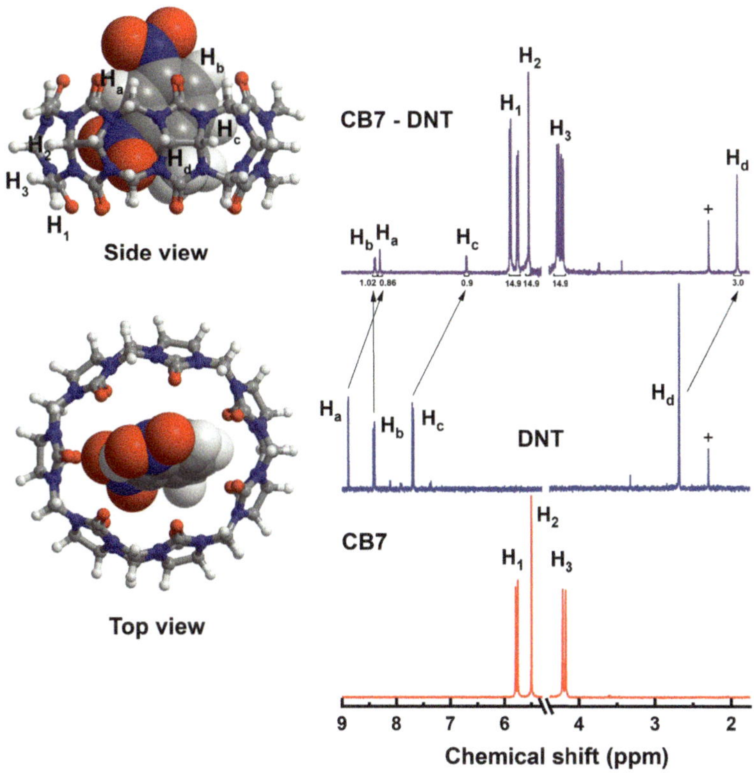

Figure 5.2 DFT molecular model of a CB7-DNT host–guest complex in water (left) and ^{1}H NMR spectra of CB7, DNT, and 1:1 CB7-DNT host–guest complexes in D_2O (right). The impurity in the DNT sample is marked by +.

5.2.1 Supramolecular Chemistry of CB7 and DNT

While the CB*n* family is known to bind a wide range of positively charged guests, the binding with neutral guests is much rarer (see Chapter 2). The formation of 1:1 host–guest complexes between CB7 and DNT in D_2O can be verified by characteristic upfield shifts of all DNT proton signals (H_a, H_b, H_c, and H_d) in the ^{1}H NMR spectra (Fig. 5.2). When excess DNT (1.2 equivalent) is added to CB7, proton signals of both complexed (H_a, H_b, H_c, and H_d) and free ($H_{a'}$, $H_{b'}$, $H_{c'}$, and $H_{d'}$) DNT can be observed, indicating that the complexation-dissociation dynamics follows the slow exchange regime on the NMR timescale at 298 K (Fig. 5.3).

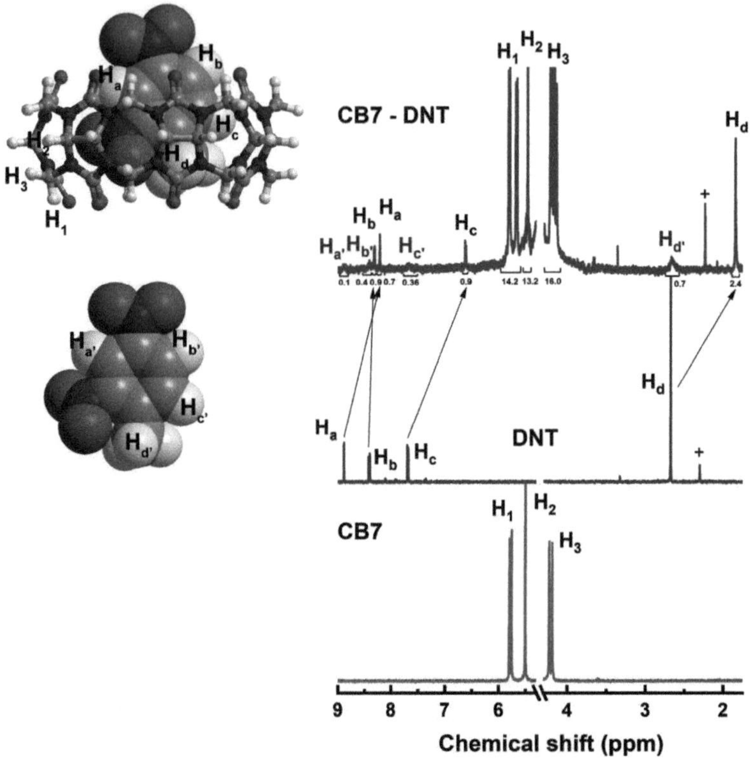

Figure 5.3 DFT molecular model of a CB7-DNT host–guest complex and a free DNT molecule in water (left) and ^1H NMR spectra of CB7, DNT, and 1:1 CB7-DNT host–guest complexes with excess DNT in D_2O (right). The impurity in the DNT sample is marked by +.

After host-guest complexation, the change in chemical shift ($\Delta\delta$) for H_d/H_c (−0.86 ppm/−1.1 ppm) of DNT is greater than that for H_b/H_a (−0.12 ppm/−0.7 ppm), indicating that H_d/H_c are deep inside the CB7 cavity but H_b/H_a are closer to the carbonyl portals. Due to the anisotropic binding of DNT, the CB7 protons desymmetrise and split the corresponding NMR signals in the CB7-DNT complex. The host–guest binding geometry derived from NMR is supported by the energy-minimised molecular model of the complex in water calculated at CPCM/wB97XD/6-311+G** level of theory (Fig. 5.2). Note that the NO_2 group at the 4-position of DNT is sticking out through the CB7 portal, thus blocking the potential interaction of one of the CB7 portals with Au NPs. Nonetheless, the other portal can still bind to the surface of Au NPs for the subsequent SERS studies.

The CB7-DNT system can be characterised by UV-Vis titration of 10 μM DNT with up to 4–5 equivalents of CB7 in order to obtain a binding constant of (1.6 ± 0.2) × 10^6 M^{-1} by assuming a 1:1 binding model (Fig. 5.4a). Isothermal titration calorimetry (ITC) can also be used to determine the thermodynamics of the CB7-DNT binding. The binding constant of (4.94 ± 0.34) × 10^5 M^{-1} obtained from ITC is consistent with UV-Vis titration within methodological error, verifying the suggested 1:1 binding mode (Fig. 5.4b). Note that it is difficult to obtain a binding constant from NMR titration in this case because of the relatively low water solubility of DNT and the relatively high aqueous binding constant of CB7-DNT. In addition, apparent reaction heat can be obtained by integrating the calorimetric traces in the ITC data. The results show that the binding is enthalpically driven ($\Delta H°$ = −11.37 kcal mol^{-1}), indicating the displacement of high-energy water molecules by DNT is the major driving force for the CB7-DNT complexation (see Section 2.4.2) [4]. It is likely that the dispersion interactions between the hydrophobic cavity of CB7 and the highly polarisable nitroaromatics contribute to the formation of host–guest complexes [5, 6]. Furthermore, negative entropic contributions ($\Delta S°$ = −0.012 kcal mol^{-1} K^{-1} or $T\Delta S°$ = −3.61 kcal mol^{-1}) reveal the formation of a tight and conformationally restricted inclusion complex, which matches with the deformed, ellipsoidal shape of CB7 in the energy-minimised molecular model (Fig. 5.2).

(a)

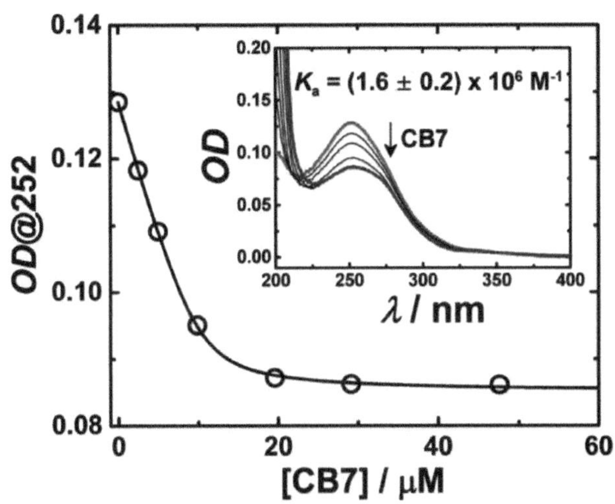

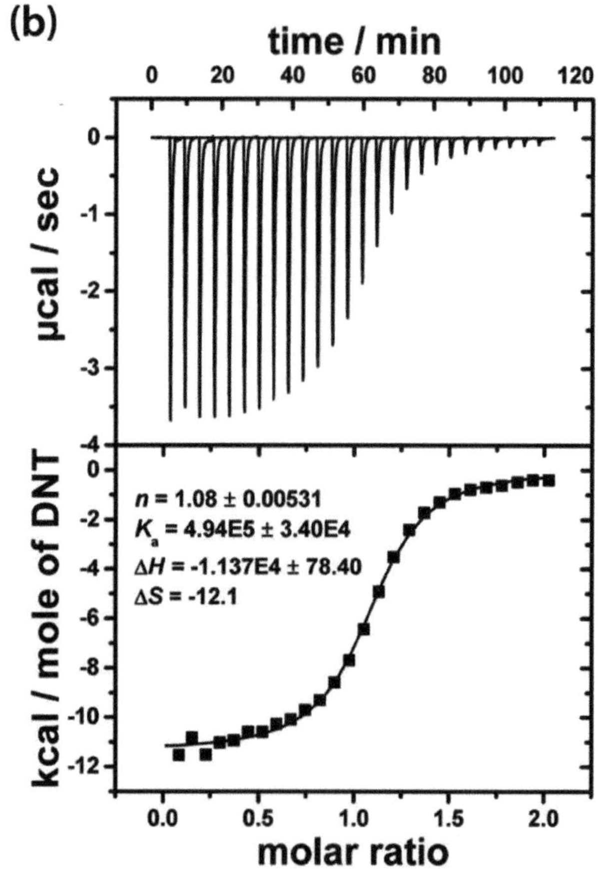

Figure 5.4 (a) UV-Vis titration of 10 μM DNT with CB7. Inset: UV-Vis spectra of CB7-DNT. (b) ITC data of CB7-DNT.

5.2.2 Raman Spectroscopy of CB7-DNT Host–Guest Complexes

CB7 crystals are characterised by two major peaks in the Raman spectrum at 444 cm^{-1} and 833 cm^{-1} (Fig. 5.5a), which are assigned to the ring scissor and ring deformation modes [7]. DNT crystals are characterised by a strong Raman peak corresponding to symmetric

NO_2 stretching vibration at 1348 cm^{-1}, and two peaks attributed to asymmetric NO_2 stretching vibration and aromatic NO_2 conjugation

(a)

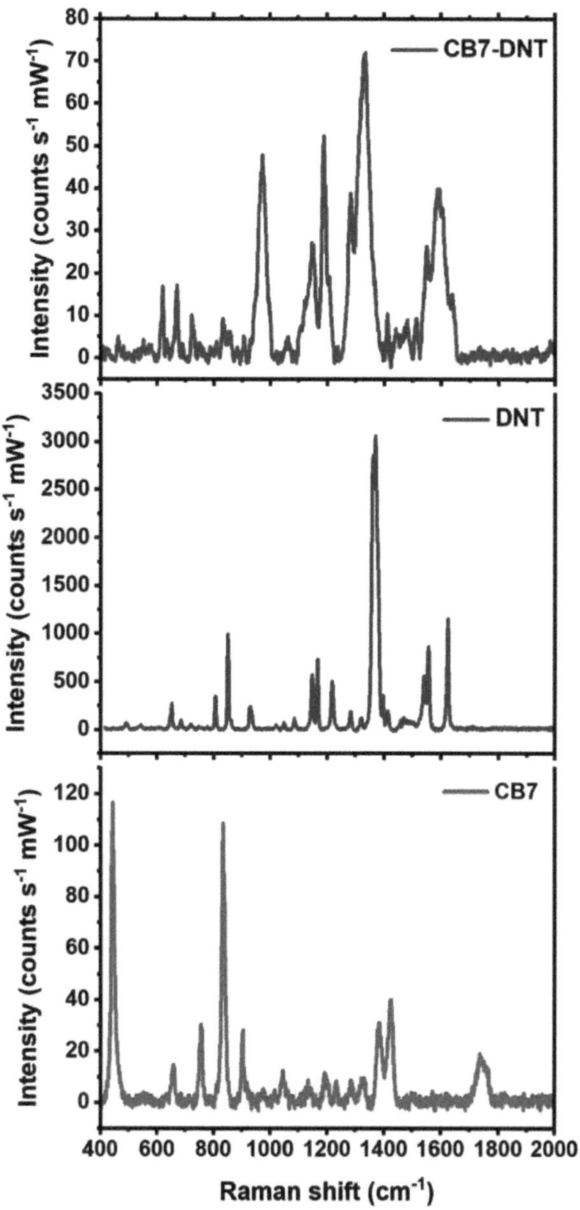

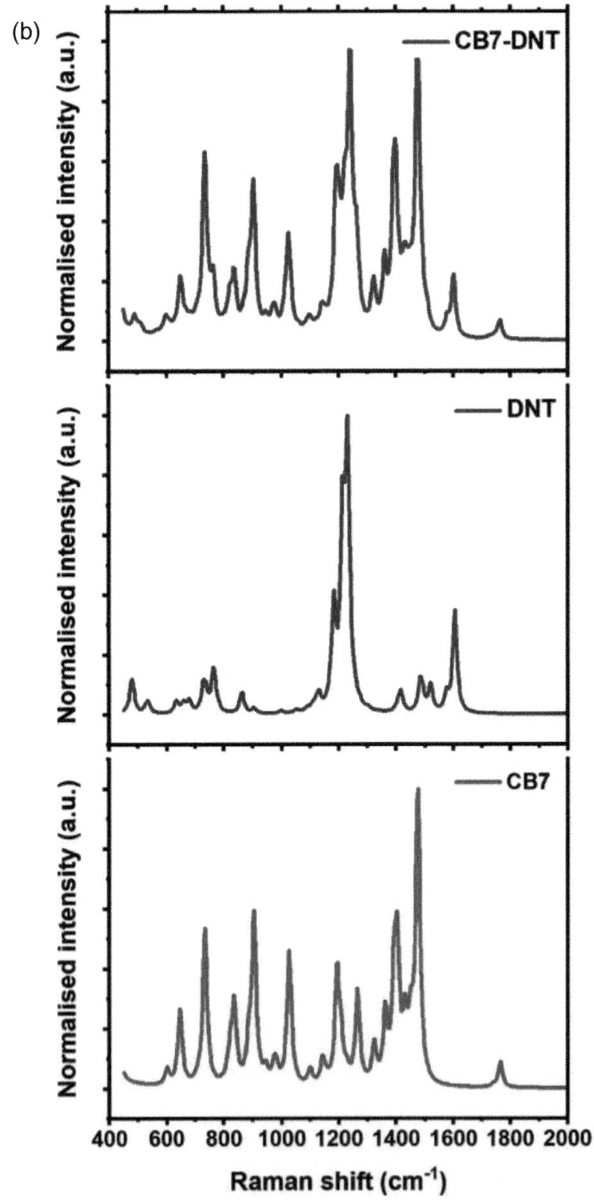

Figure 5.5 (a) Experimental Raman spectra of CB7 crystals, DNT crystals, and 1:1 CB7-DNT host–guest complexes of 2 mM in water. (b) Calculated Raman spectra of CB7, DNT, and 1:1 CB7-DNT host–guest complexes in gas phase at the HF/3-21G level of theory.

at 1544 cm^{-1} and 1611 cm^{-1}, respectively [8]. Note that there are slight shifts in positions of the Raman signals of the 1:1 CB7-DNT host–guest complexes in aqueous solution when compared to those of CB7 and DNT crystals. The DNT signals dominate in the Raman spectrum of the CB7-DNT complexes, owing to the fact that DNT is more Raman-active than CB7. The Raman spectra of CB7, DNT, and the CB7-DNT complexes modelled in the gas phase at the HF/3-21G level of theory show general consistency with the experimental results (Fig. 5.5b). The calculated Raman spectrum of the CB7-DNT complex is approximately equal to the sum of that of CB7 and DNT despite slight shifts in some peaks.

5.2.3 Sensitivity of the SERS Nanosensor for Nitroexplosives

Macrocycle-functionalised nanosensors for SERS detection of DNT can be prepared by pre-mixing CB7 and DNT solutions with 1:1 stoichiometry and adding it into a 40 nm citrate-capped Au NP solution. CB7 mediates the formation of the nanoaggregates as evidenced by the red to purple colour change of the Au NP solution and the formation of ~1.0 nm plasmonic nanojunctions shown in the transmission electron microscopy (TEM) images (Figs. 5.1 and 5.6).

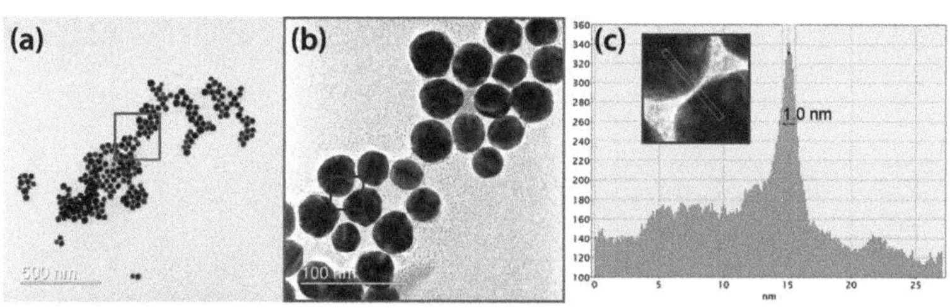

Figure 5.6 (a–c) Stepwise zoom-in TEM images of 40 nm Au NPs after adding CB7-DNT complexes.

When DNT is added to Au NPs without CB7, there is no aggregation of Au NPs and thus no DNT signal can be detected (Fig. 5.7a). In contrast, the characteristic Raman peak of DNT at 1334 cm^{-1} can be clearly observed in the presence of CB7, illustrating the successful detection of DNT by the CB7-Au NP nanosensors and the importance of the macrocycles on DNT sensing. It should be noted

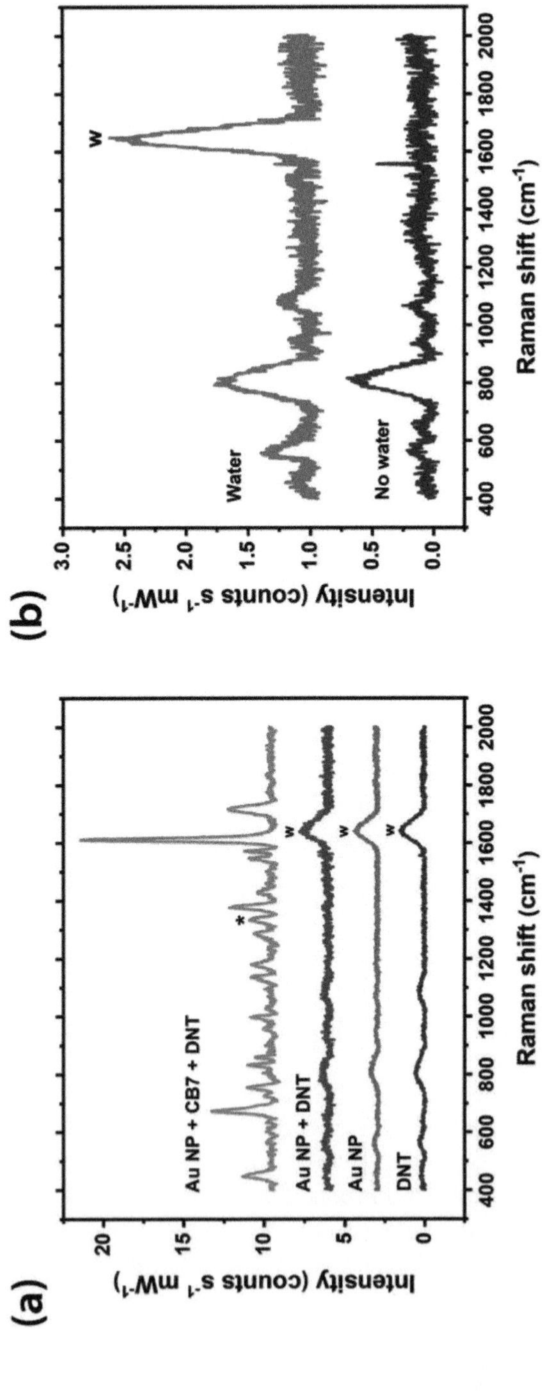
(a)
Intensity (counts s-1 mW-1)
Au NP + CB7 + DNT
Au NP + DNT
Au NP
DNT
w
*
Raman shift (cm-1)
(b)
Intensity (counts s-1 mW-1)
Water
No water
w
Raman shift (cm-1)

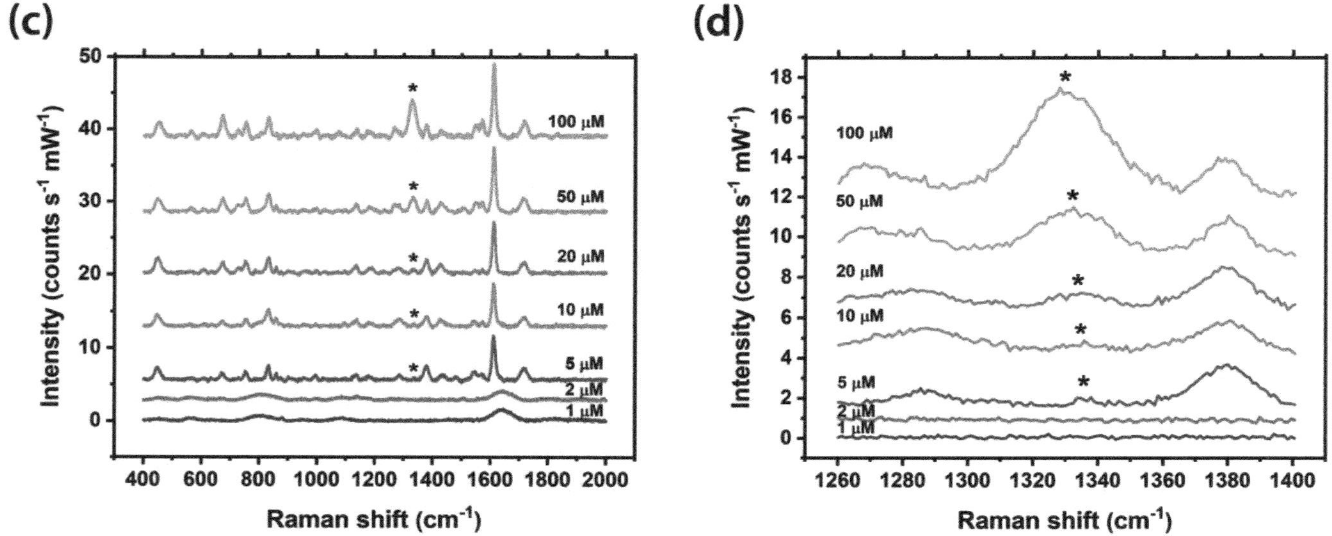

Figure 5.7 (a) Raman and SERS spectra of DNT with or without CB7. Main Raman peak of DNT is marked by *. (b) Raman spectrum of water in a custom-made sample holder. The H-O-H bending peak of liquid water is marked by w. (c) Full-range and (d) zoom-in SERS spectra of the 1:1 CB7-DNT complexes from 1 to 100 μM.

that the broad peaks at ~1640 cm^{-1} in the Raman spectra of DNT and Au NPs are attributed to the H-O-H bending mode of water (Fig. 5.7b). In order to determine the appropriate concentration of CB7 for the nanosensors, sequential dilution of the 1:1 CB7-DNT host–guest complexes should be performed before adding into the Au NP solution for SERS sensing (Fig. 5.7c,d). The characteristic signals of DNT are readily observable for concentrations down to ~10 μM. At very low concentrations of the CB7-DNT complexes, no SERS signals can be observed as there is an insufficient amount of CB7 to trigger the self-assembly of Au NPs.

The detection limit of the CB7-Au NP nanosensors for nitroexplosives can be tested by performing SERS measurements of DNT at 10 μM CB7 (Fig. 5.8a,b). As the aggregation kinetics is determined by the ratio of CB:Au NPs (see Sections 3.3.1 and 3.4.1), a constant CB7 concentration is used to ensure the formation of reproducible SERS substrates [2]. The main peak of DNT attributed to symmetric NO_2 stretching vibration can be clearly observed when the concentration is down to 2 μM (Fig. 5.8c). Logarithmic fittings are commonly used to reveal a correlation between SERS intensity and concentration of nitroexplosives. The results show that the detection limit is ~1 μM with a strong correlation (R^2 ~ 0.97) between the SERS intensity and log concentration of DNT from 0.5 to 10 μM (Fig. 5.8d), which is comparable to other solution-based SERS sensors [9, 10].

Note that the CB7-DNT complexes alone cannot mediate the aggregation of Au NPs as one of the CB7 portals is blocked by the bulky nitro group of DNT, which can be evidenced from the energy-minimised molecular model and the ^{1}H NMR spectra of the CB7-DNT complexes (Fig. 5.2). The aggregation of Au NPs is mainly mediated by the empty CB7 present at the dynamic equilibrium of the host–guest complexation. Nonetheless, the CB7-DNT complexes can still remain in close proximity to the plasmonic hotspots by binding to the surface of Au NPs via the accessible portal (Fig. 5.8a). This idea is supported by the relatively large error bar and a slight departure from linearity in the 20 μM DNT sample (Fig. 5.8d), where the concentration of empty CB7 is relatively lower than the other samples due to a shift in equilibrium position caused by the presence of excess DNT.

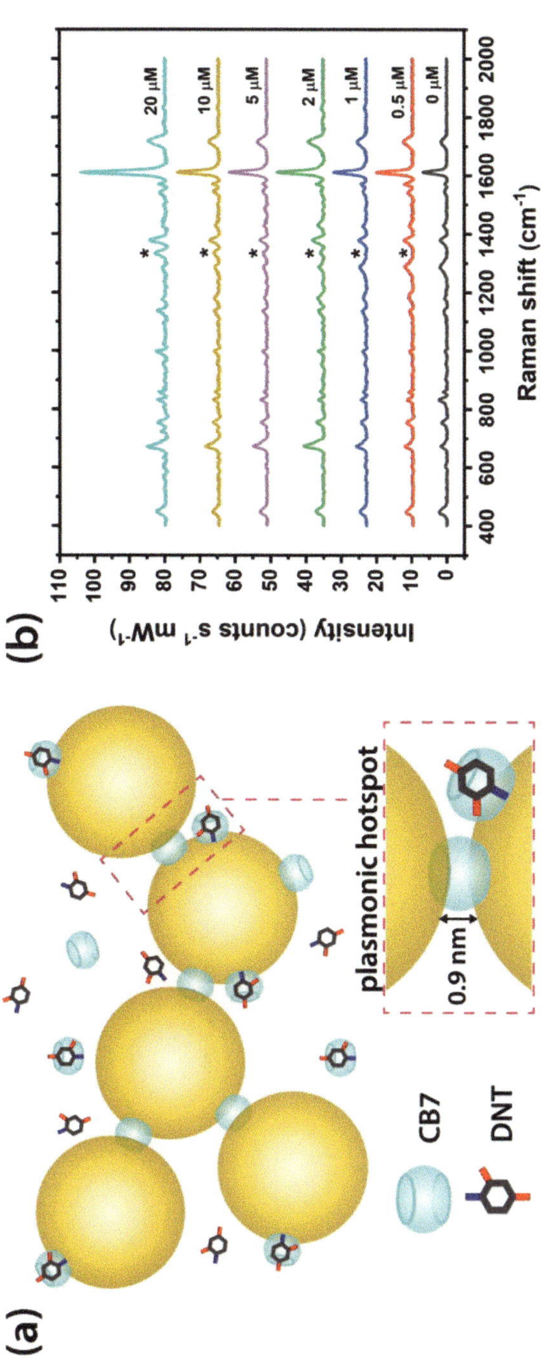
(a)
CB7
DNT
plasmonic hotspot
0.9 nm
(b)
Intensity (counts s⁻¹ mW⁻¹)
Raman shift (cm⁻¹)
20 μM
10 μM
5 μM
2 μM
1 μM
0.5 μM
0 μM

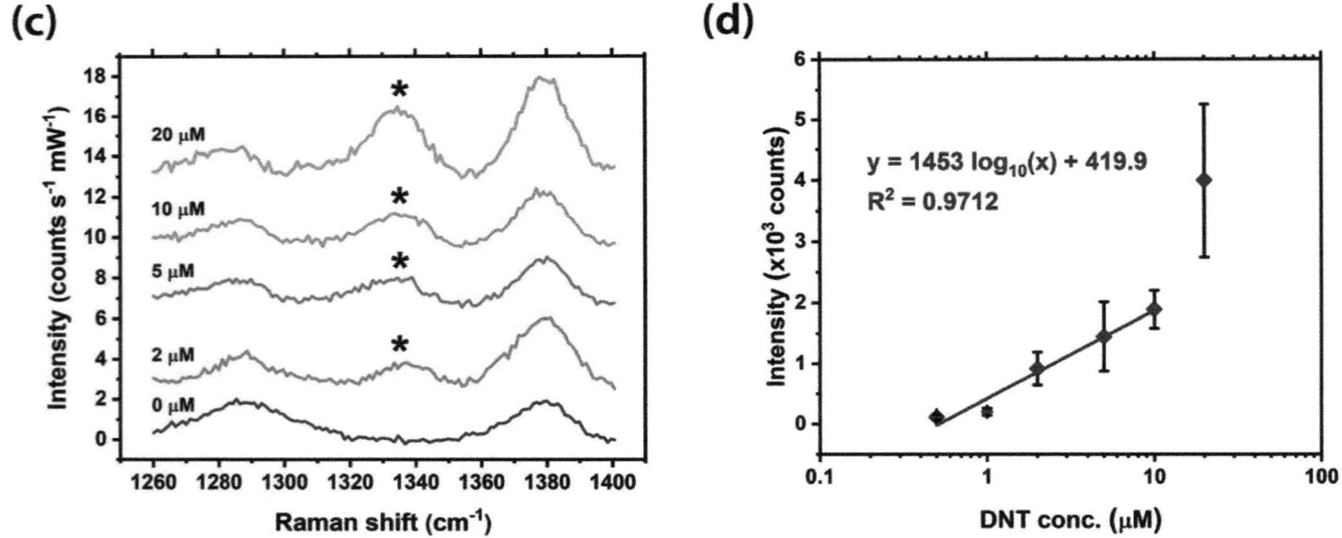

Figure 5.8 (a) Schematic illustration of the precise plasmonic hotspots within the CB7-Au NP nanosensor for DNT detection (not to scale). (b) Full-range and (c) zoom-in SERS spectra of DNT with concentrations from 0 to 20 μM. Main Raman peak of DNT is marked by *. (d) Corresponding plot of SERS intensity of the main DNT peak in (b) against its concentration.

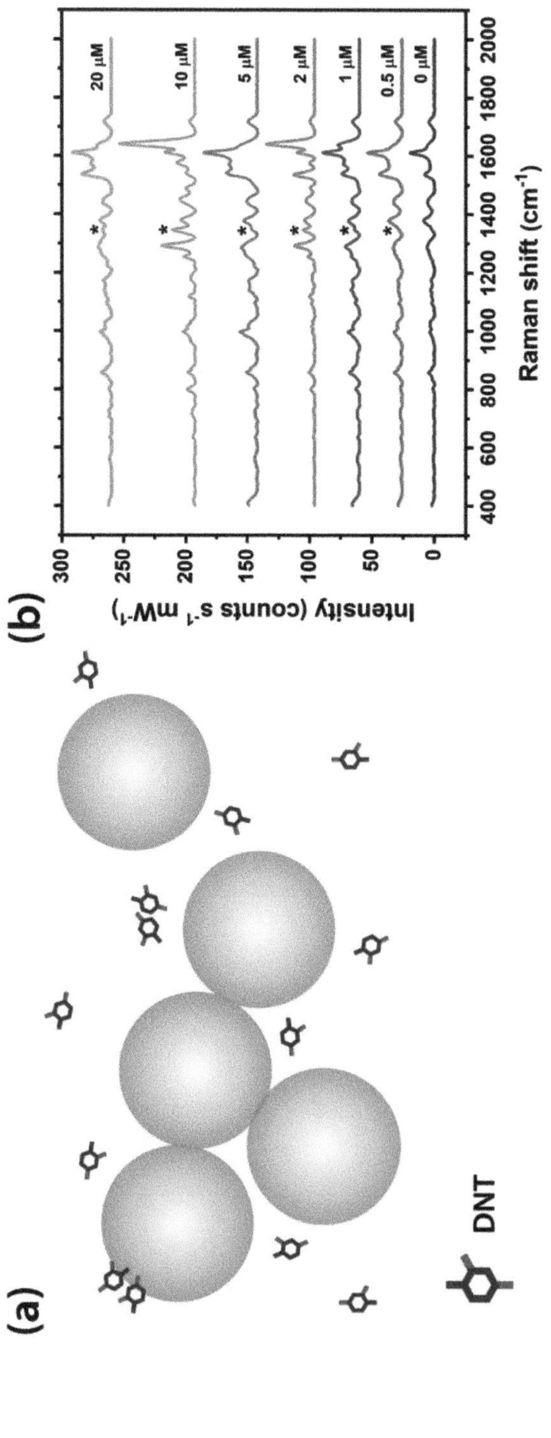
(a)
DNT
(b)
Intensity (counts s-1 mW-1)
Raman shift (cm-1)
20 μM
10 μM
5 μM
2 μM
1 μM
0.5 μM
0 μM
300
250
200
150
100
50
0
400
600
800
1000
1200
1400
1600
1800
2000

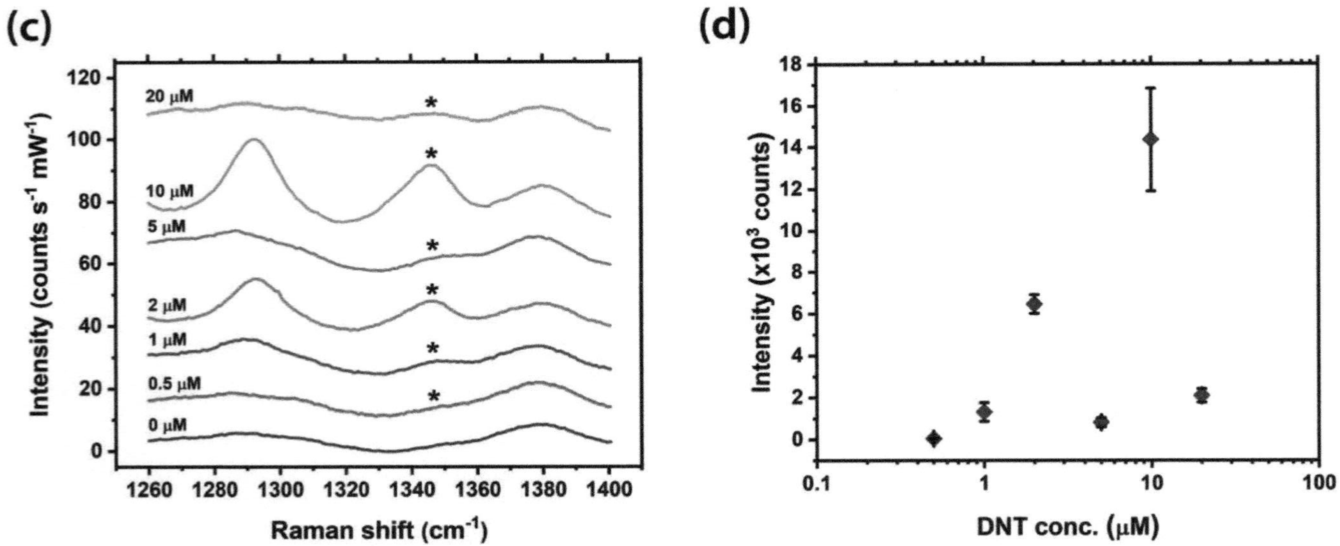

Figure 5.9 (a) Schematic illustration of the random plasmonic hotspots within the NaCl-Au NP sensor for DNT detection (not to scale). (b) Full-range and (c) zoom-in SERS spectra of DNT with concentrations from 0 to 20 μM. Main Raman peak of DNT is marked by *. (d) Corresponding plot of SERS intensity of the main DNT peak in (b) against its concentration.

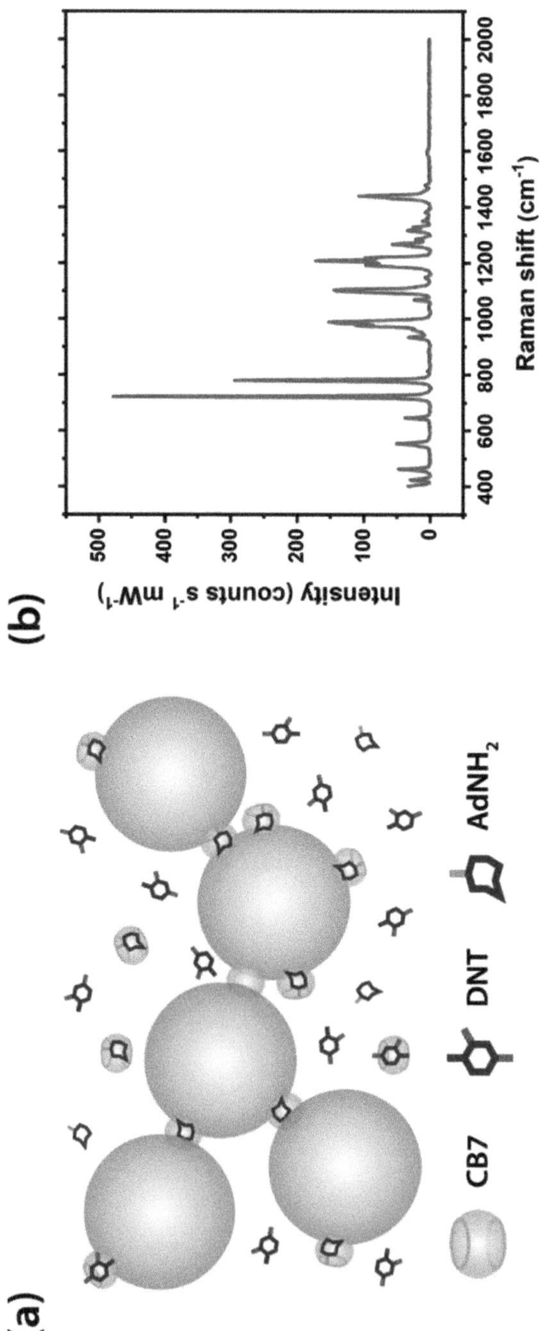
(a)
CB7
DNT
AdNH2
(b)
Intensity (counts s-1 mW-1)
500
400
300
200
100
0
400
600
800
1000
1200
1400
1600
1800
2000
Raman shift (cm-1)

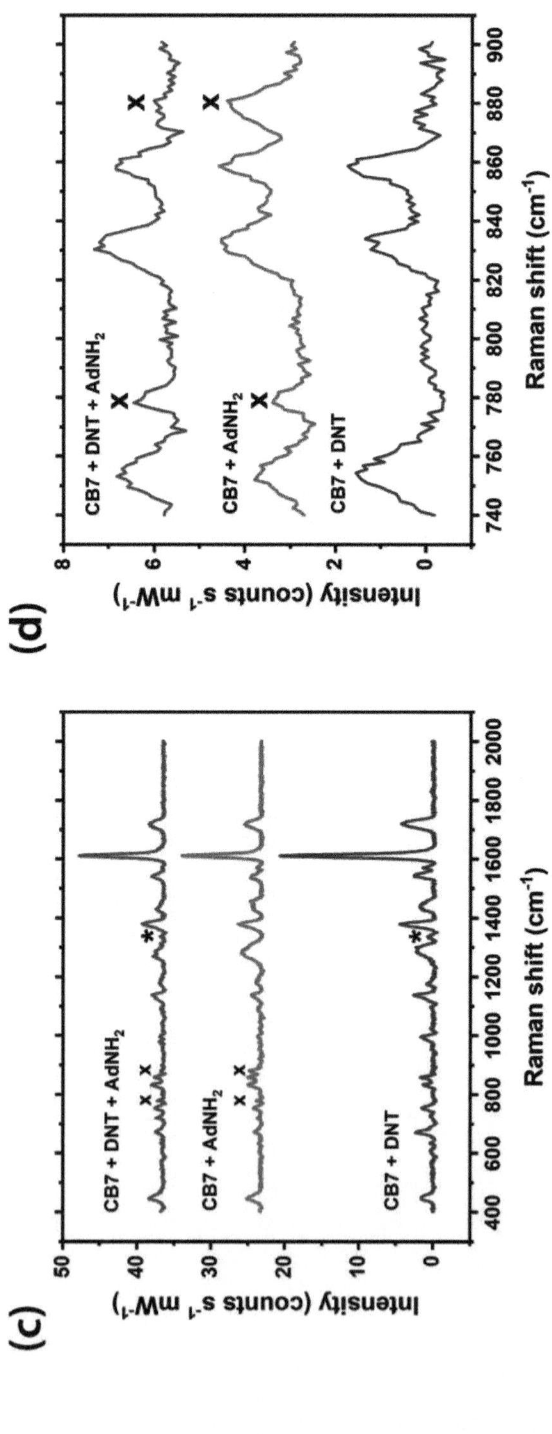
(c)
Intensity (counts s-1 mW-1)
CB7 + DNT + AdNH2
CB7 + AdNH2
CB7 + DNT
Raman shift (cm-1)
(d)
Intensity (counts s-1 mW-1)
CB7 + DNT + AdNH2
CB7 + AdNH2
CB7 + DNT
Raman shift (cm-1)

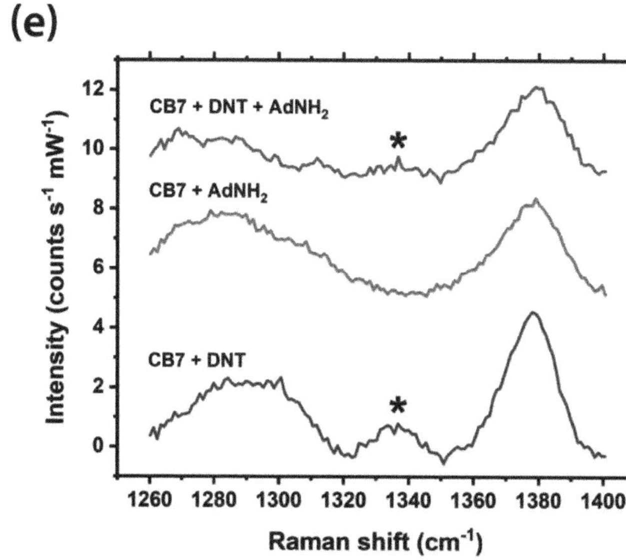

Figure 5.10 (a) Schematic illustration of the precise plasmonic hotspots within the CB7-Au NP nanosensor for DNT and $AdNH_2$ detection (not to scale). (b) Raman spectrum of $AdNH_2$ powder. (c) Full-range and (d, e) zoom-in SERS spectra showing competitive formation of CB7-DNT and CB7-$AdNH_2$ host–guest complexes. Main Raman peaks of DNT and $AdNH_2$ are marked by * and x, respectively.

The macrocycle plays two major roles in enhancing the signals of the nitroaromatic explosives. Firstly, the rigid structure of CB7 defines precise nanojunctions of 0.9 nm between the Au NPs, leading to the formation of very strong and localised plasmonic hotspots. This phenomenon can be supported by a control experiment with NaCl-Au NP sensors where 0.1 M NaCl is used to aggregate Au NPs (Fig. 5.9a). Although DNT can be detected by the NaCl-Au NP SERS sensors, the characteristic peak of DNT is significantly broader, and, more importantly, a correlation between the SERS intensity and concentration is lacking, owing to the formation of random plasmonic hotspots in uncontrolled aggregation (Fig. 5.9b–d). Secondly, the surface-bound CB7 can encapsulate DNT inside its cavity, leading to an enriched concentration on the surface of Au NPs while still in close proximity to the plasmonic hotspots. This is particularly important for analyte molecules such as nitroaromatic compounds that have low intrinsic affinities to the Au surface. The surface-enrichment effect of CB7 can be verified by a competitive binding experiment with 1-adamantylamine ($AdNH_2$), a strongly binding guest of CB7 (Fig. 5.10). As a result, there is a 67% reduction in the SERS signal of the main peak of DNT in the presence of equimolar (10 μM) $AdNH_2$ which means $AdNH_2$ has effectively displaced DNT from the CB7 cavity (Fig. 5.10c,e).

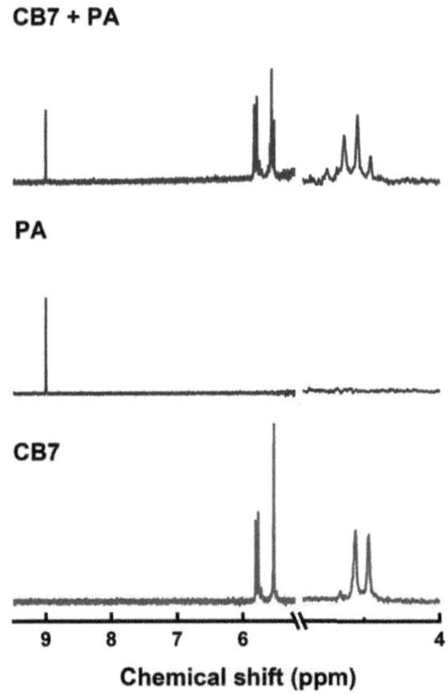

Figure 5.11 ^{1}H NMR spectra of CB7, PA and 1:1 CB7 and PA in D_2O.

5.2.4 Selectivity of the SERS Nanosensor for Nitroexplosives

The selectivity of the CB7-Au NP nanosensor can be demonstrated with a structurally similar nitroexplosive, picric acid (PA), which is too bulky to fit inside the CB7 cavity. Different from the case of CB7-DNT, no characteristic upfield shift of the PA proton signals can be observed in ^{1}H NMR when CB7 and PA were mixed in D_2O with 1:1 stoichiometry (Fig. 5.11). This verifies the absence of host–guest complexation between CB7 and PA. In addition, the overall negative charge and high water solubility of PA disfavour its binding with CB7. Despite being ca. 3.5 times more Raman-active than DNT, PA did not exhibit any observable SERS signals in the presence of 10 μM CB7 (Fig. 5.12), thus illustrating the significance of host–guest complexation in the macrocycle-functionalised nanosensors.

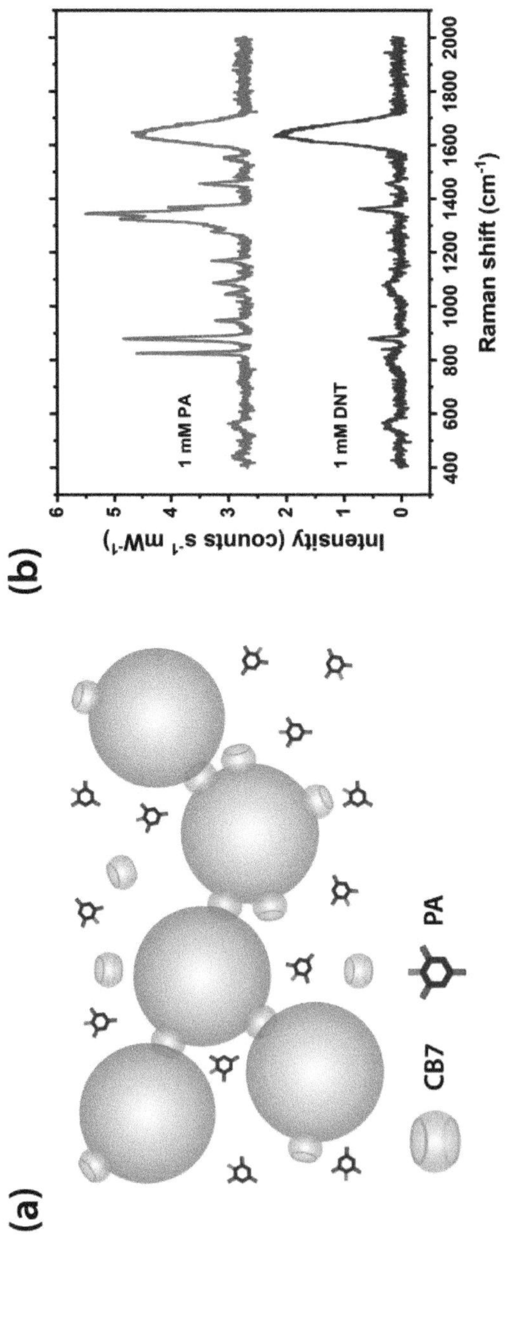
(a)
CB7
PA
(b)
1 mM PA
1 mM DNT
Intensity (counts s^{-1} mW^{-1})
Raman shift (cm^{-1})

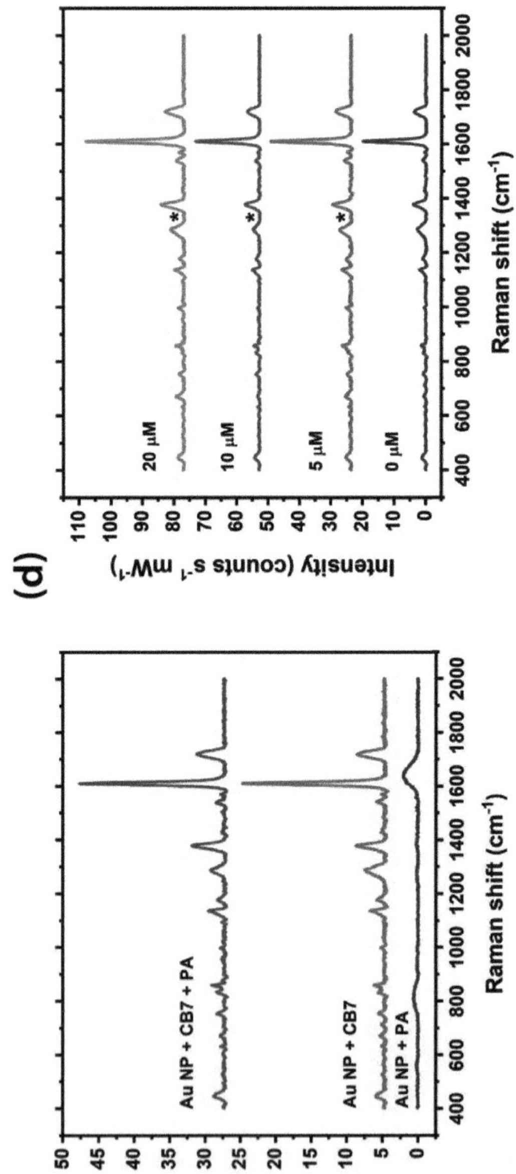

(c)
Intensity (counts s-1 mW-1)
Raman shift (cm-1)
Au NP + CB7 + PA
Au NP + CB7
Au NP + PA
(d)
Intensity (counts s-1 mW-1)
Raman shift (cm-1)
20 μM
10 μM
5 μM
0 μM

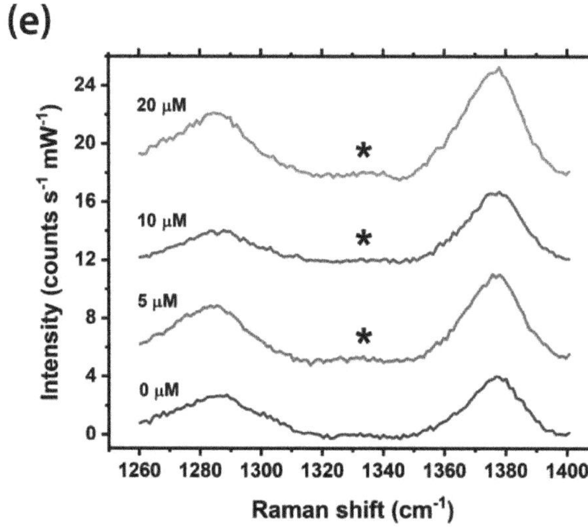

Figure 5.12 (a) Schematic illustration of the precise plasmonic hotspots within the CB7-Au NP nanosensor in the absence of host–guest complex between CB7 and PA (not to scale). (b) Raman spectra of 1 mM PA and 1 mM DNT. (c) SERS spectra of PA with or without CB7. (d) Full-range and (e) zoom-in SERS spectra of PA with concentrations from 0 to 20 μM.

5.2.5 Tolerance of the SERS Nanosensor for Nitroexplosives

The tolerance of the CB7-Au NP nanosensor can be demonstrated by performing SERS detection of nitroexplosives in the presence of strong Au NP ligands such as 4-mercaptobenezoic acid (4-MBA) and 4-mercaptophenylboronic acid (4-MPBA), as to mimic strong binding environmental impurities in field applications. While 4-MBA can strongly bind to Au NPs, it cannot be detected at 1 μM in the absence of CB7 (Fig. 5.13c). In contrast, the 4-MBA signals can be picked up by the CB7-Au NP nanosensor at the same concentration, illustrating the importance of the macrocycle in the formation of precise plasmonic hotspots (Fig. 5.13a). The characteristic peaks of DNT and 4-MBA can be clearly observed in multiplexed SERS sensing of 10 μM DNT and 1 μM 4-MBA (Fig. 5.13d–f). Note that the nanosensor is also capable to simultaneously detect DNT and 4-MPBA, thus reflecting its robustness for trace nitroexplosives sensing. Furthermore, it is also possible to eliminate the interference from various background contaminants by extracting the Raman shift information, unlike the classical on/off-type sensors which are more prone to false-positive results.

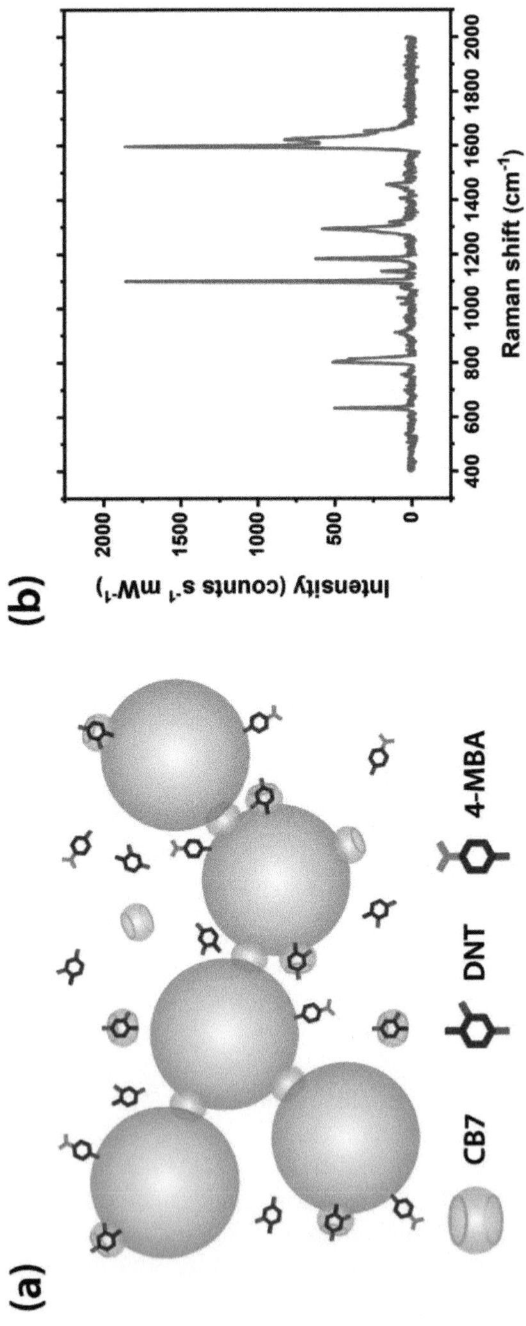
(a)
CB7
DNT
4-MBA
(b)
Intensity (counts s-1 mW-1)
2000
1500
1000
500
0
400 600 800 1000 1200 1400 1600 1800 2000
Raman shift (cm-1)

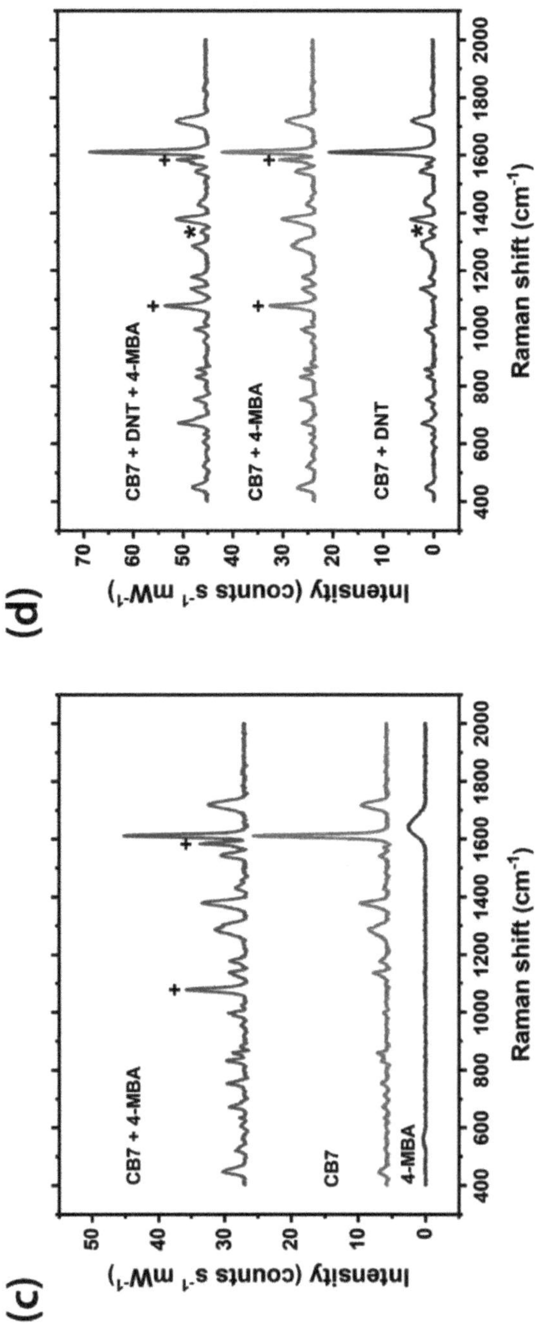
(c)
CB7 + 4-MBA
CB7
4-MBA
Intensity (counts s-1 mW-1)
Raman shift (cm-1)
(d)
CB7 + DNT + 4-MBA
CB7 + 4-MBA
CB7 + DNT
Intensity (counts s-1 mW-1)
Raman shift (cm-1)

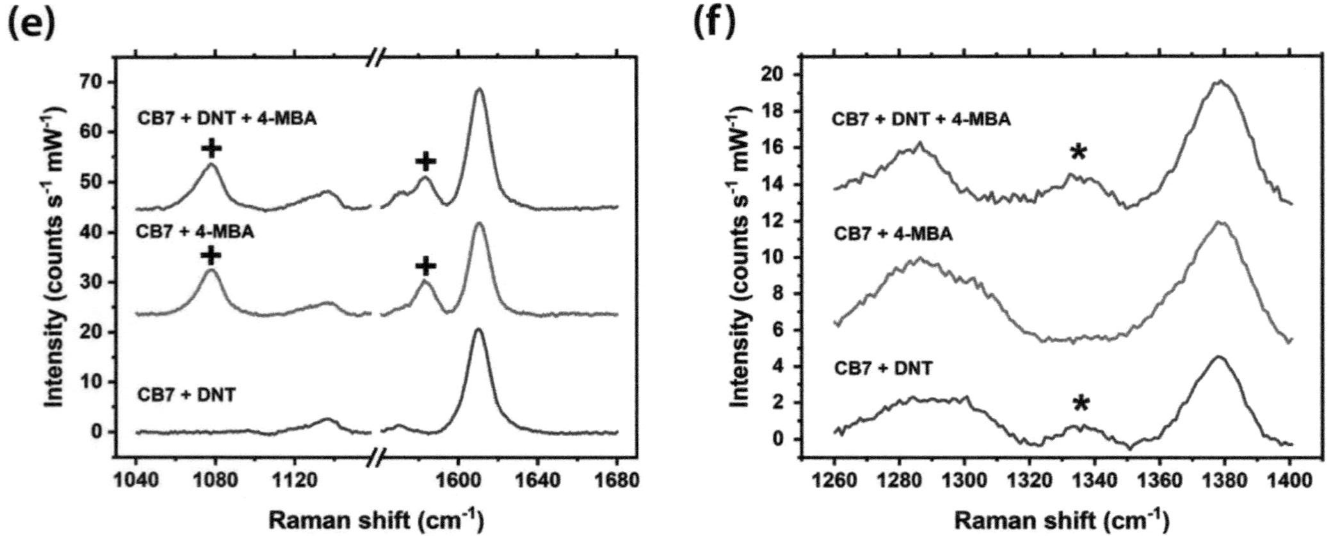

Figure 5.13 (a) Schematic illustration of the precise plasmonic hotspots within CB7-Au NP nanosensor for DNT and 4-MBA detection (not to scale). (b) Raman spectrum of 4-MBA powder. (c) SERS spectra of 4-MBA with or without CB7. (d) Full-range and (e–f) zoom-in SERS spectra of 10 μM DNT with 1 μM 4-MBA. Main Raman peaks of DNT and 4-MBA are marked by * and +, respectively.

5.2.6 Summary

- The CB7-Au NP nanosensor can detect poly-substituted nitroaromatic explosive DNT by forming neutral host–guest complexes between CB7 and DNT in aqueous solution.
- The key binding parameters of the CB7-DNT complexes can be quantified using NMR, UV-Vis titrations, and ITC measurements and supported by DFT molecular models.
- The nanosensor allows SERS quantification of DNT with highly reproducible signals via formation of precise plasmonic hotspots, meanwhile offering rapid response, high reproducibility, and signal linearity.
- The nanosensor is highly selective against a structurally similar nitroexplosive, picric acid (PA), which is too bulky to be encapsulated within the CB7 cavity.
- The surface-enrichment effect of CB7 can be demonstrated by a competitive binding experiment with 1-adamantylamine ($AdNH_2$).
- The nanosensor is tolerant against the presence of organic contaminants, 4-mercaptobenezoic acid (4-MBA) and 4-mercaptophenylboronic acid (4-MPBA), that strongly bind to Au NPs.

5.3 CB7-QD Nanosensors for Nitroexplosives

Apart from mediating the aggregation of Au NPs, the rigid macrocycle (i.e., CB7) can also functionalise fluorescent quantum dots (QDs) to form nanosensors for detection of nitroexplosives via aqueous self-assembly (Fig. 5.14). CB7 has a relatively high water solubility. Meanwhile, CdTe has bright emission and good stability in aqueous solution with tunable absorption and emission properties. Herein, this case study will discuss the interaction of CB7 with the surface of metal chalcogenide CdTe, QDs, energy transfer between different QDs, as well as the detection of DNT and NB using the CB7-QD nanosensor [11].

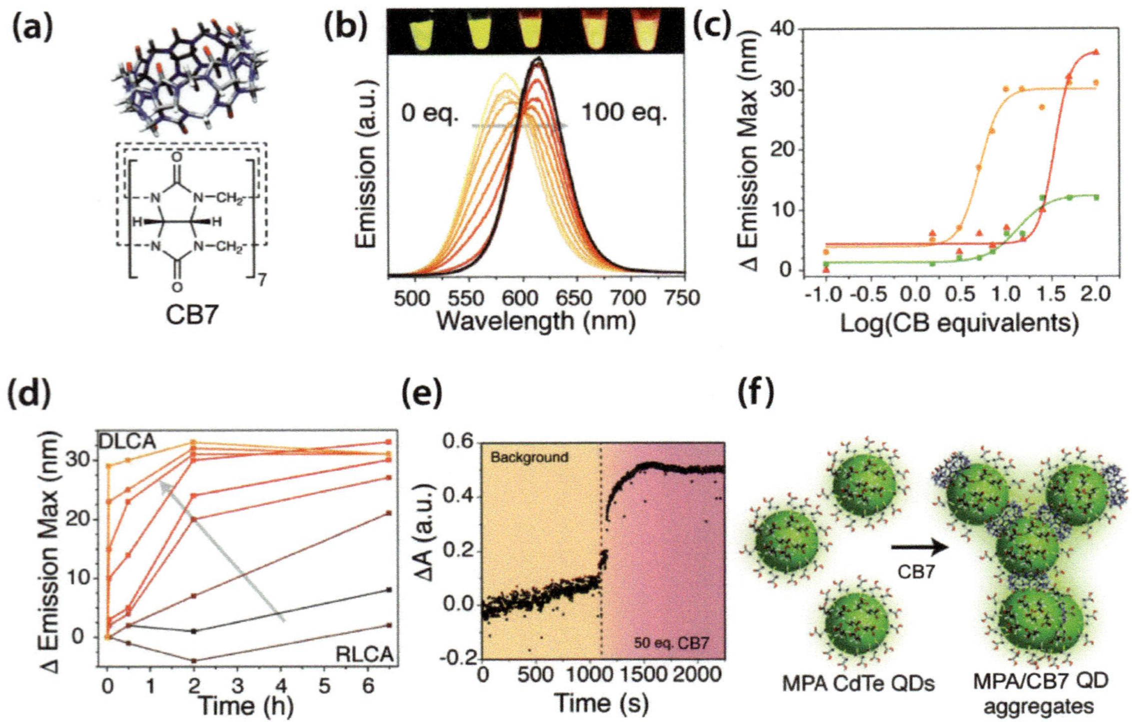

Figure 5.14 (a) Molecular structure of CB7. (b) Emission spectra of CB7-QD. Inset: QD_O with 1, 5, 10, 25, and 50 eq. of CB7 under UV illumination (from left to right). (c) Change in maximum emission wavelength of the QD_R, QD_O, and QD_G at 15 μM each, with increasing CB7 equivalents. (d) Change in maximum emission wavelength for QD_O over time with 0, 0.1, 1.5, 3, 5, 7, 10, 15, and 50 eq. of CB7, respectively. (e) Change in optical extinction of the QD_R when 50 eq. of CB7 is added. (f) Schematic illustration of ligand exchange on the surface of QDs by CB7.

5.3.1 Functionalisation of CdTe QDs by CB7

CB7-QD nanoaggregates can be formed by simply mixing the 3-mercaptopropionic acid (MPA) capped CdTe QDs (red, orange, or green) at different concentrations of CB7 (Fig. 5.14a) in deionised water. When the ratio of CB7:QD is above 1:1 (15 μM QD), brightening of the orange QDs (QD_O) and a red-shift of 10–35 nm in the emission wavelength can be observed under UV illumination, due to the formation of CB7-QD nanoaggregates and energy transfer from the slightly smaller to larger QDs within the same population (Fig. 5.14b,c). The variation in the inflection points of the maximum emission wavelengths is owing to the surface area of particles, sample heterogeneity, and coverage of the thiol ligands. Similar aggregation effects can also be observed with MPA-capped CdSe/ZnS QDs, but with less red-shifting, due to the more homogeneous particle distribution. Note that the CB7-QD nanoaggregates grow over time with a rate of growth depending on the amount of CB7 added to the solution (Fig. 5.14d). At low CB7 equivalents (1–5 eq.), it can take several hours for the larger nanoaggregates to grow, resembling the reaction limited colloidal aggregation (RLCA) [12, 13]. On the other side, the nanoaggregates formed rapidly and started to flocculate out of the solution within 12 h through diffusion limited colloidal aggregation (DLCA) at high CB7 equivalents (>30 eq.), i.e., when CB7 is in large excess. A higher temporal resolution of the latter regime can be obtained by measuring the absorption or scattering of light of the samples (Fig. 5.14e). When CB7 is added to the red QDs (QD_R), there is an instant increase in optical extinction, followed by a slight drop over time due to the formation and precipitation of the larger nanoaggregates.

The size of the larger nanoaggregates can be measured by dynamic light scattering (DLS), which shows the small nanoaggregates are 10–100 nm and the larger ones can grow over 1000 nm with more CB7 equivalents (Fig. 5.15a). It is suggested that CB7 can exchange with the MPA ligands on the surface of QDs, causing aggregation of the QDs by binding to one or both portals of CB7 in a similar fashion to the other NPs, meanwhile enabling energy transfer (Fig. 5.14f). The rate of aggregation and the size of the clusters are determined by the total concentration and ratio of CB7:QD. When the ratio is greater than 7:1–15:1, there is excess CB7 for an ideally packed surface. Note that the residual signals of MPA are still visible in washed aggregates formed with 50 eq. of CB7, revealing that only partial displacement of the stronger thiol ligands by weaker CB7 has taken place, as shown by the Fourier transform infrared spectroscopy (FTIR) data (Fig. 5.15b). The C=O stretching of CB7 at 1713 cm^{-1} blue-shifts by ~10 cm^{-1} and broadens slightly in the CB7-QD nanoaggregates which indicates the interaction between CB7 and QDs is through the electron-rich carbonyl portals of CB7.

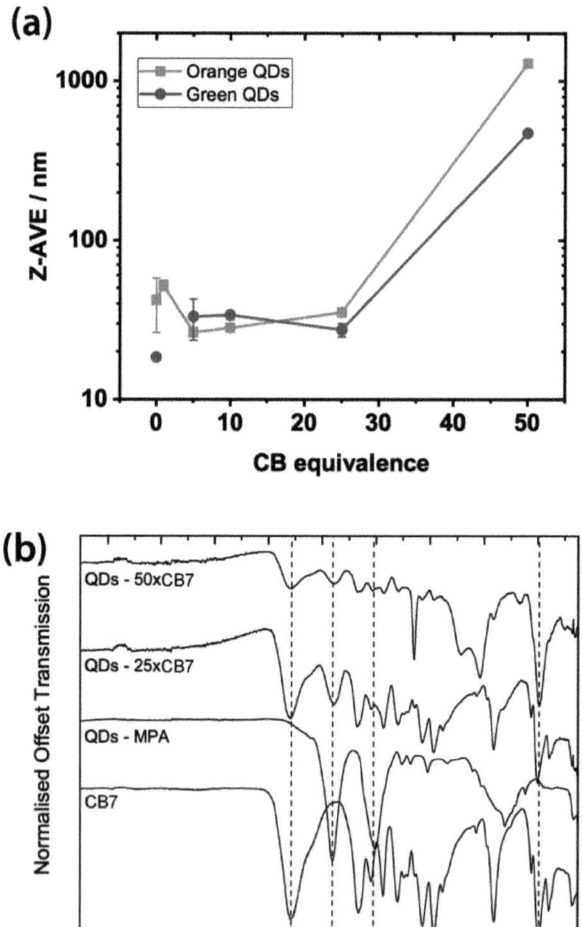

Figure 5.15 (a) DLS data for QD_G and QD_O with increasing eq. of CB7. (b) FTIR data of QD_O, CB7, and QD_O with 25 or 50 eq. of CB7. Major peaks of CB7 and MPA are indicated by the dotted lines.

In addition, NMR can be used to study the CB7-QD interaction. Several molecular environments of MPA between 2.3 and 3.0 ppm, arising from the CH_2–CH_2 region, can be seen in the ^1H NMR spectra of 15 μM QD_R in D_2O with increasing CB7 concentrations (Fig. 5.16a). Also, two pairs of signals at 2.45/2.75 ppm and 2.35/2.57 ppm can be identified in 2D COSY experiments (data not shown). It can be deduced from ^{13}C NMR spectroscopy and mass spectrometry that the larger sharp triplet pair (2.45/2.75 ppm) originated from the oxidised MPA dimer in solution while the second sharp pair (2.35/2.57 ppm) might come from transient binding of free MPA on the surface of QDs (data not shown) [14]. Furthermore, there is a featureless and considerably broadened third pair of signals (~2.6/2.9 ppm) which is not present in the 2D COSY analysis (data not shown). Compared to the NMR signals of free MPA, the third pair of signals is downfield shifted, and it is postulated that the MPA molecules tightly bound to the surface of QDs via the Cd-S bonds give rise to the signals. The extreme broadness can be attributed to the slow motion of QD-MPA conjugates and the multiple binding sites for MPA on the surface of QDs.

When CB7 is added to the QD solution, three further downfield-shifted signals can be observed at 4.10, 5.40, and 5.65 ppm, respectively (Fig. 5.16a). The heavily broadened signals indicate there is interaction between CB7 and QDs. Note that the CB7 signals broaden further with increasing CB7 concentration, as more CB7 molecules are driven to the surface of QDs, thus increasing the size of the nanoaggregates (trapping more CB7) and slowing diffusion on NMR timescale.

5.3.2 Energy Transfer within the CB7-QD Nanoaggregates

The steady-state energy transfer within the CB7–QD nanoaggregates can be studied by mixing QDs of green and red emissions where the emission spectrum of the QD_G overlaps with the absorption of the QD_R. When the QDs are mixed in solution without CB7, the deconvoluted emissions of QDs show that their positions and intensities features are similar to the individual spectral features before mixing (Fig. 5.16b). Meanwhile, the red-shift of the peaks and, more importantly, quenching of the QD_G emission and brightening of

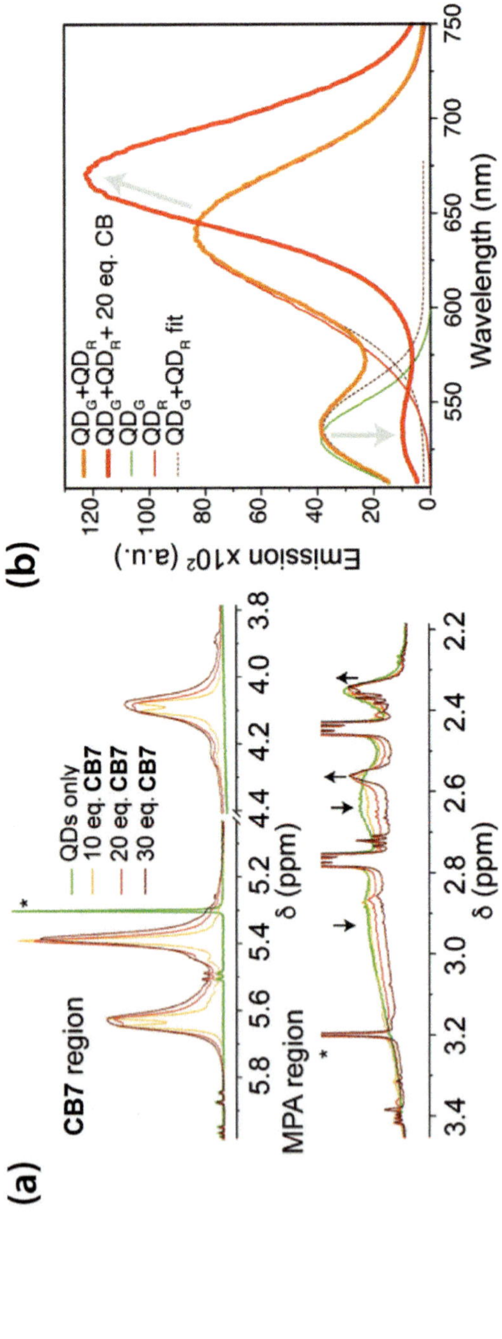
(a)
CB7 region
QDs only
10 eq. CB7
20 eq. CB7
30 eq. CB7
5.8 5.6 5.4 5.2 4.4 4.2 4.0 3.8
δ (ppm)
MPA region
3.4 3.2 3.0 2.8 2.6 2.4 2.2
δ (ppm)
(b)
QD_G+QD_R
QD_G+QD_R+ 20 eq. CB
QD_G
QD_R
QD_G+QD_R fit
Emission x10^2 (a.u.)
0 20 40 60 80 100 120
550 600 650 700 750
Wavelength (nm)

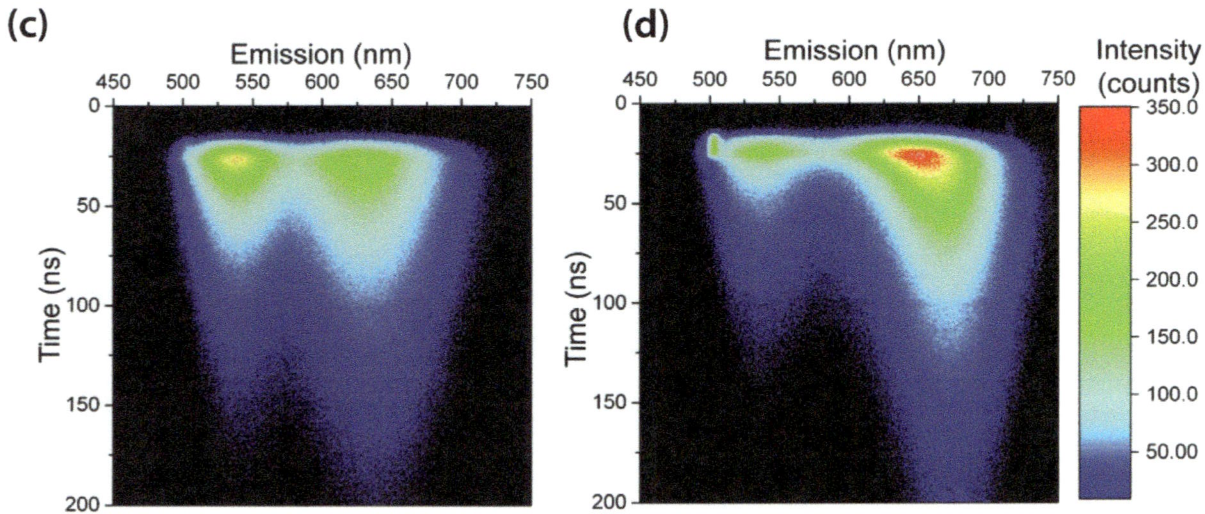

Figure 5.16 (a) ^{1}H NMR spectra of 15 μM QD_R in D_2O. Changes with increasing concentrations of CB7 are indicated by arrows while the solvent residues are marked by *. (b) Emission spectra for a mixture of QD_G and QD_R with or without 20 eq. of CB7. (c, d) 2D mapping of emission wavelength and decay for a mixture of QDs (c) without and (d) with CB7. The bright signal at 20 ns/500 nm is due to scattering of excitation light.

the QD_R emission can be observed when 20 eq. of CB7 is added to the QD mixture, indicating efficient energy transfer is caused by mixing and gluing of different QDs. In addition, a more complex energy cascade can also be achieved for a mixture of three QDs (red, orange, and green) in which changes in intensity and peak wavelength can be observed by performing multi-Gaussian peak fitting (Fig. 5.17).

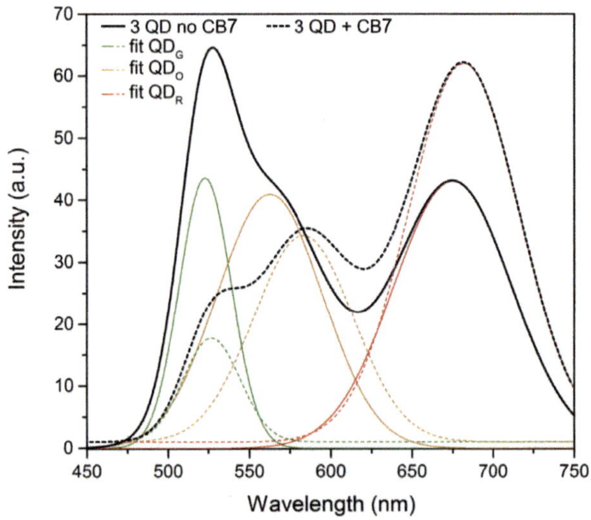

Figure 5.17 Emission spectra of three-colour QDs with or without 50 eq. of CB7 in water. Samples without CB7 are indicated by solid lines while those with CB7 are indicated by dashed lines.

The dynamic nature of the CB7-QD nanoaggregates can be probed by measuring energy transfer between a mixture of two pre-formed QD nanoaggregates (green and red) over time. In particular, 15 μM QD_G and QD_R solutions were mixed with 10 eq. of CB7, and their emissions were measured after 30 min. Note that low equivalents of CB7 should be used to make sure CB7 and QD react completely within the desired time frame and the observed effects are not caused by the presence of residual CB7 (Fig. 5.18a). There are a few possible routes for the aggregation of CB7 and two-colour QDs. When both colours of QDs were mixed at a 1:1 ratio with a total concentration of 15 μM, there was an obvious change in fluorescence over 90 min in the presence of CB7, but the effect was not very noticeable in the absence of CB7 (Fig. 5.18b). The dual-colour QD nanoaggregates can also be formed by mixing pre-formed CB7-QD_G nanoaggregates

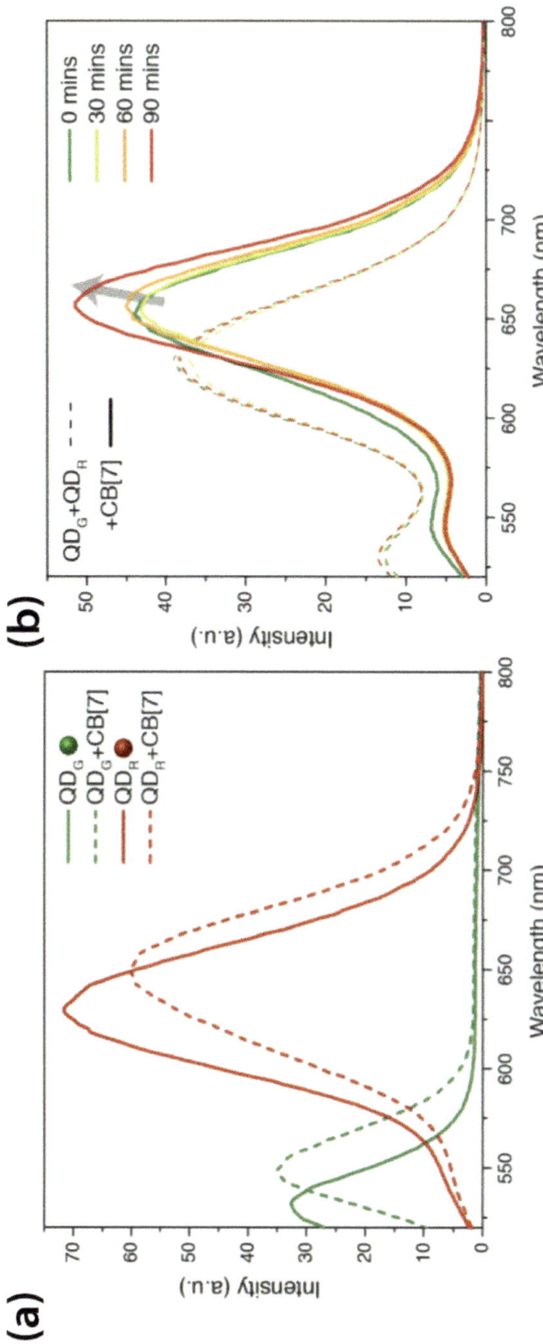
(a)
Intensity (a.u.)
Wavelength (nm)
QD$_G$
QD$_G$+CB[7]
QD$_R$
QD$_R$+CB[7]
(b)
Intensity (a.u.)
Wavelength (nm)
QD$_G$+QD$_R$
+CB[7]
0 mins
30 mins
60 mins
90 mins

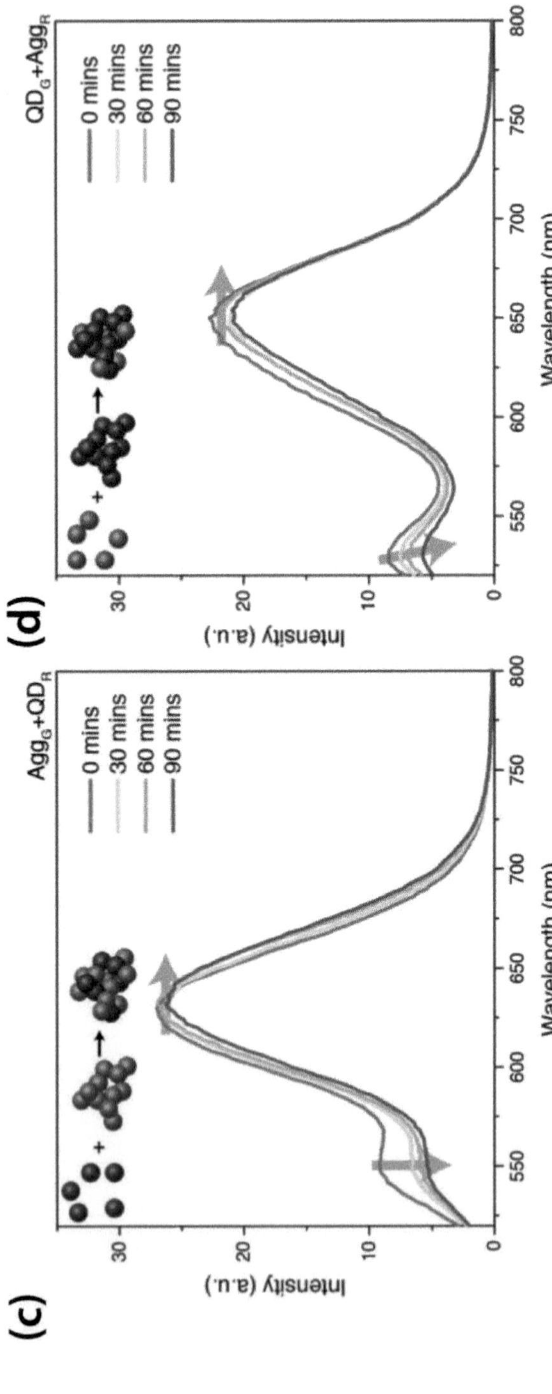
(c)
Agg_G+QD_R
0 mins
30 mins
60 mins
90 mins
Intensity (a.u.)
Wavelength (nm)
(d)
QD_G+Agg_R
0 mins
30 mins
60 mins
90 mins
Intensity (a.u.)
Wavelength (nm)

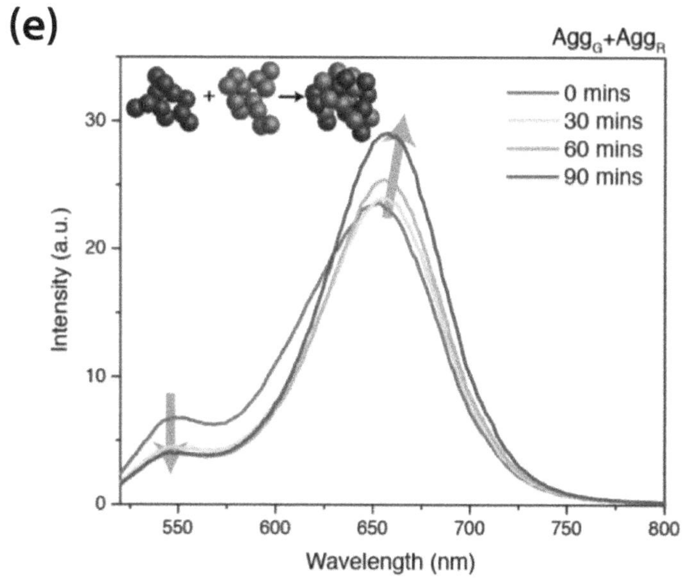

Figure 5.18 Emission spectra of (a) QD_G or QD_R with or without CB7, (b) a mixture of QD_G and QD_R with or without CB7, (c) a mixture of pre-formed CB7-QD_G nanoaggregates (Agg_G) with stand-alone QD_R, (d) a mixture of pre-formed CB7-QD_R nanoaggregates (Agg_R) with stand-alone QD_G, and (e) a mixture of pre-formed CB7-QD_G (Agg_G) and CB7-QD_R nanoaggregates (Agg_R).

(Agg_G) with stand-alone QD_R and vice versa. The addition of QD_R into the solution of Agg_G incorporates the QD_R into the nanoaggregates, resulting in a loss of green intensity and red-shifting of the QD_R (Fig. 5.16c). Similarly, the incorporation of the QD_G into the pre-formed CB7-QD_R nanoaggregates (Agg_R) causes red-shifting of the QD_G (Fig. 5.15d). In both cases, there were only slight changes in the red emissions, probably because full integration of QDs into the nanoaggregates and energy transfer of green-to-red were not achieved during the 90-min time frame. In the last scenario, the dual-colour nanoaggregates were formed by mixing Agg_G and Agg_R, which showed energy transfer of green-to-red within 90 min, due to the presence of higher CB7 concentration (Fig. 5.18e).

The energy transfer in the CB7–QD nanoaggregates can be further studied by measuring the time-resolved emission of a mixture of QD_G and QD_R with or without CB7 (Fig. 5.16c,d). The emission lifetimes were calculated for a full range of the QD_G emission peak (510–570 nm, 37 ns) and for three regions of the broader QD_R peak (blue-edge, 600–630 nm, 47 ns; centre, 630–670 nm, 52 ns; and red-edge, 670–730 nm, 57 ns). Notably, QD_G has a third, short component (2.6 ns) after aggregation, which moves it from bi- to tri-exponential, thus suggesting a fast energy transfer. For QD_R, the blue-edge and central region become bi-exponential, with a decrease in the average lifetime of the blue-edge (35 ns). Meanwhile, the central region has both shortened and lengthened components (30 ns/95 ns) owing to the convolution of the green-to-red and red-to-red energy transfer processes. The red-edge region remains approximately the same as they are the final acceptor of energy transfer. Nevertheless, it shows an increased rise time which indicates it originated from an energy transfer process instead of direct laser excitation.

5.3.3 Fluorescent Nanosensors for Nitroexplosives

Apart from CB7-Au NP nanoaggregates, macrocycle-functionalised nanosensors for the detection of nitroexplosives, DNT and NB, can be prepared by adding the pre-mixed CB7-DNT and CB7-NB solution into the QD solution, respectively. The formation of the CB7-DNT-QD_O and CB7-NB-QD_G nanoaggregates can be evidenced by the red-shifting of the fluorescence emission band (Figs. 5.19a,b and 5.20). In addition, the quenching of QD_O increases by approximately fourfold in the presence of CB7 and DNT while there is a clear enhancement for NB with CB7 despite the very poor fit due to an outliner at 1 μM in the absence of CB7 (Fig. 5.21). Note that there is a reduced red-shift of the QDs at high concentrations of DNT (0.35 and 0.7 mM, >1 DNT per CB7), indicating one of the CB7 portals is blocked by the nitroexplosives and thus full integration of the nanoaggregates may be hindered (Fig. 5.19c). This phenomenon can be confirmed by mixing two-colour QDs (Q_G and Q_R) and 0.35 mM DNT in the presence of CB7, where increased fluorescence quenching by PET and increased green emission by reduced energy transfer (ET) from Q_G to Q_D are observed (Fig. 5.22).

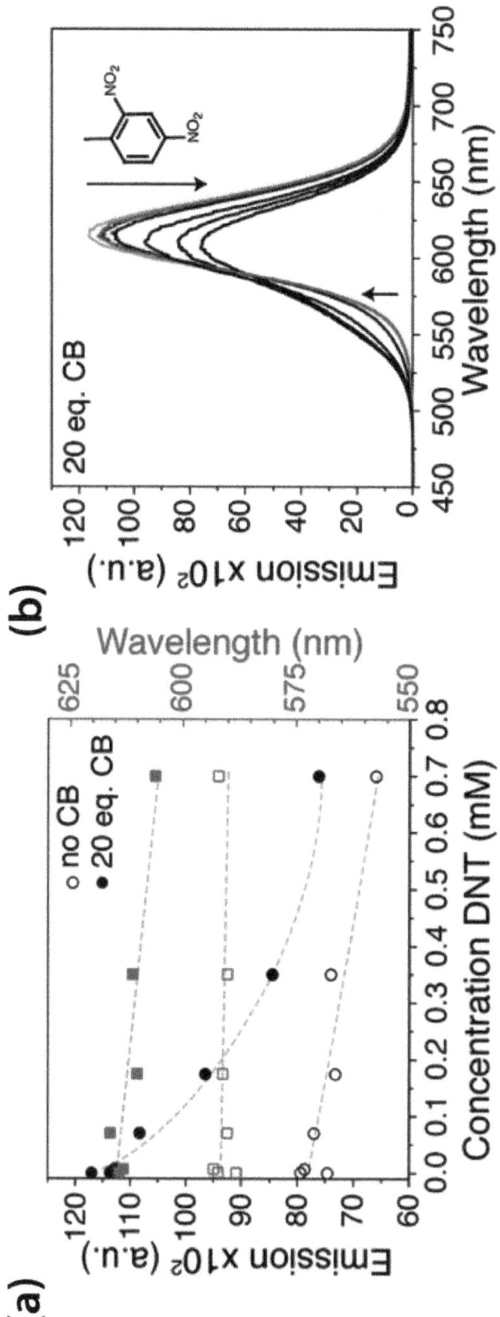
(a)
(b)
Emission x10² (a.u.)
Concentration DNT (mM)
Wavelength (nm)
o no CB
• 20 eq. CB
20 eq. CB
NO_2
Wavelength (nm)

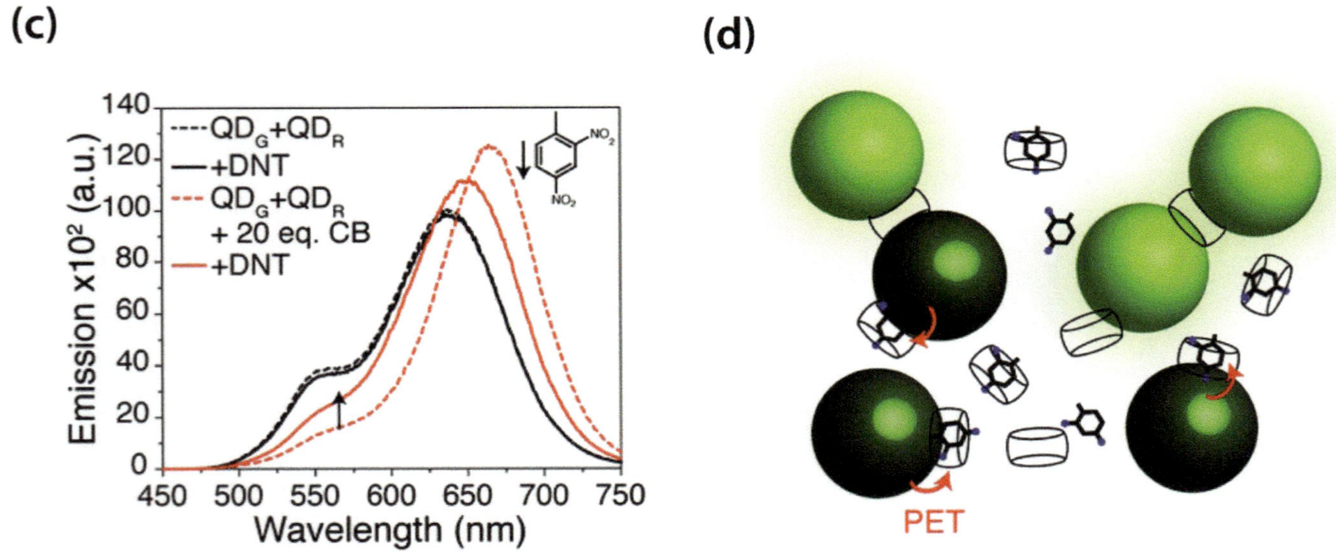

Figure 5.19 (a) Emission spectra of 0–0.7 mM DNT with or without CB7. The raw spectra in the presence of CB7 shown in (b). Black circles and red squares correspond to change in intensity and emission maximum wavelength, respectively. (c) Emission spectra of a dual-QD system in the presence of 0.35 mM DNT with or without CB7. (d) Schematic illustration of the CB7-QD$_G$ nanosensor for the DNT detection. Photoinduced electron transfer from QD to DNT is indicated by arrows. MPA is not shown for clarity.

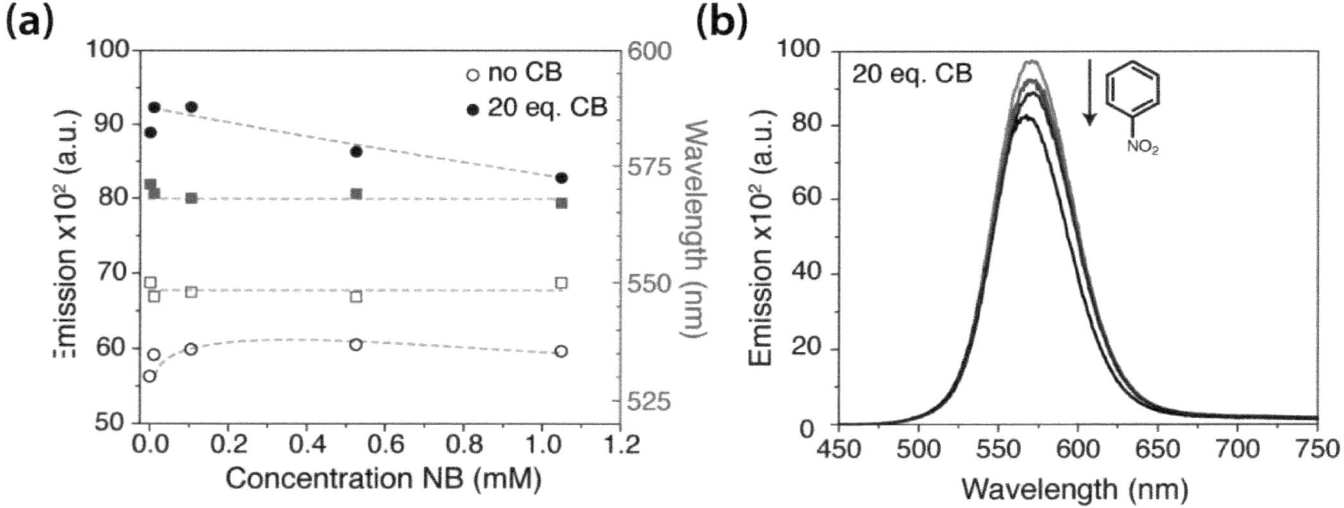

Figure 5.20 (a) Emission spectra of 0–1.05 mM NB with or without CB7. The raw spectra in the presence of CB7 shown in (b). Black circles and red squares correspond to change in intensity and emission maximum wavelength, respectively.

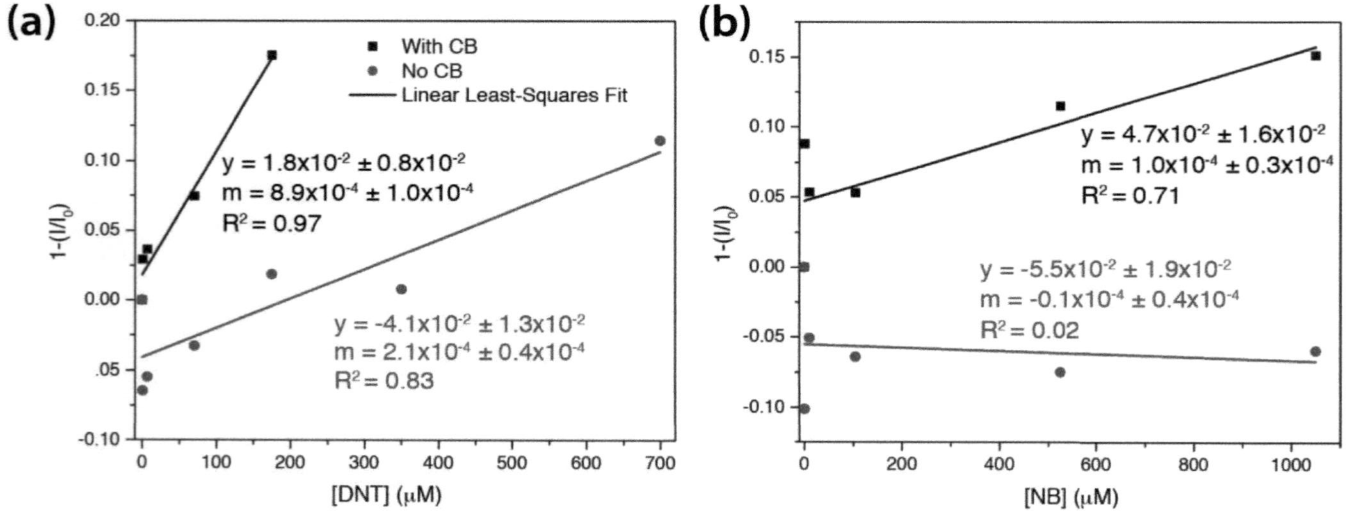

Figure 5.21 Linear fittings of (a) DNT and (b) NB titrations.

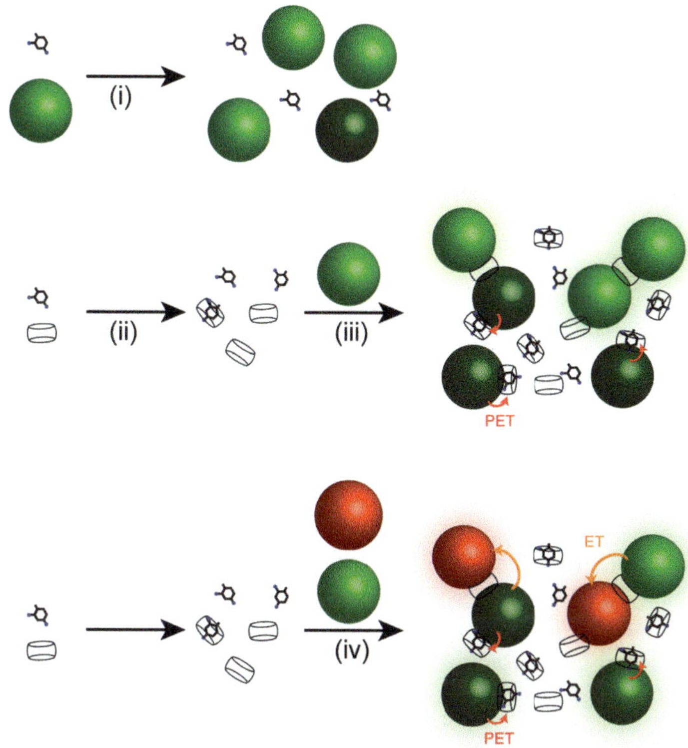

Figure 5.22 Schematic illustration of the CB7-QD_G and CB7-QD_G-QD_R nanosensor for DNT detection, with PET from QD_G to DNT and ET from QD_G to QD_R.

When an electron-withdrawing molecule is encapsulated by a macrocycle, the fluorescence of the system can be perturbed, enabling it to act as a reporter of binding [15]. The significance of the electron-withdrawing nature in fluorescence nanosensing can be verified by the lack of quenching of toluene in a control experiment where a very little change is observed upon toluene addition (Fig. 5.23). Although non-specific trapping of the electron-withdrawing molecules in the nanoaggregates can also cause quenching, ^{1}H NMR, titration studies, and DFT simulation have verified the CB7-DNT host–guest complexation (see Section 5.2.1). Note that photoinduced electron transfer (PET), the probable quenching mechanism in the nanosensor, works on short distances, suggesting tight binding of DNT molecules on the surface of QD mediated by CB7.

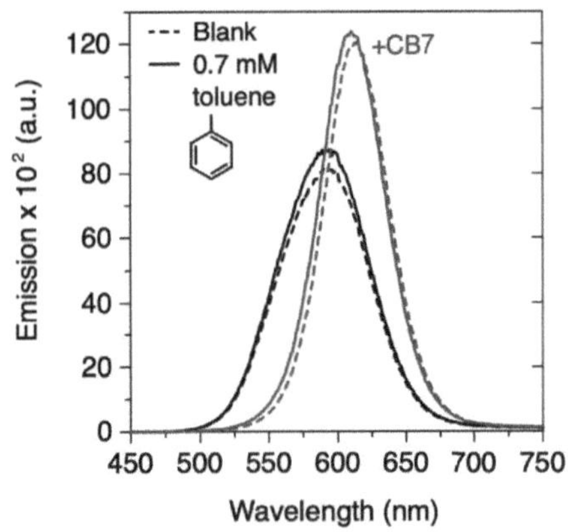

Figure 5.23 Emission spectra for QD_O in the presence or absence of 0.7 mM toluene, with or without CB7.

5.3.4 Summary

- The CB7-QD nanosensor can detect DNT and NB by forming host–guest complexes in aqueous solution.
- Self-assembly of the nanoaggregates in aqueous media is mediated by CB7 via portal binding, with their size being dependent on the ratio of CB:QD.
- The aggregation mechanism can be verified by NMR, IR, and photoluminescence measurements, in which CB7-QD nanoaggregates self-assembled from 10–20 equivalents of CB per QD are colloidally stable for solution sensing.
- Energy transfer cascades occur within the nanoaggregates from smaller to larger QDs in two- and three-colour QD systems.
- The CB7 bound to QDs can remain supramolecularly active which can increase quenching of QDs by several-fold via a surface enrichment mechanism.

References

1. Fang, Y., Seong, N.-H. and Dlott, D. D. (2008). Measurement of the distribution of site enhancements in surface-enhanced Raman scattering, *Science*, **321**, pp. 388–392.

2. Ben-Jaber, S., Peveler, W. J., Quesada-Cabrera, R., Cortes, E., Sotelo-Vazquez, C., Abdul-Karim, N., Maier, S. A. and Parkin, I. P. (2016). Photo-induced enhanced Raman spectroscopy for universal ultra-trace detection of explosives, pollutants and biomolecules, *Nat. Commun.*, **7**, 12189.
3. Chio, W.-I. K., Peveler, W. J., Assaf, K. I., Moorthy, S., Nau, W. M., Parkin, I. P., Olivo, M. and Lee, T.-C. (2019). Selective detection of nitroexplosives using molecular recognition within self-assembled plasmonic nanojunctions, *J. Phys. Chem. C*, **123**, pp. 15769–15776.
4. Biedermann, F., Uzunova, V. D., Scherman, O. A., Nau, W. M. and De Simone, A. (2012). Release of high-energy water as an essential driving force for high-affinity binding of cucurbit[*n*]urils, *J. Am. Chem. Soc.*, **134**, pp. 15318–15323.
5. Biedermann, F., Nau, W. M. and Schneider, H.-J. (2014). The hydrophobic effect revisited – studies with supramolecular complexes imply high-energy water as a noncovalent driving force, *Angew. Chem. Int. Ed.*, **53**, pp. 11158–11171.
6. Liu, T. and Schneider, H.-J. (2002). Additivity and quantification of dispersive interactions – from cyclopropyl to nitro groups: measurements on porphyrin derivatives, *Angew. Chem. Int. Ed.*, **41**, pp. 1368–1370.
7. Mahajan, S., Lee, T.-C., Biedermann, F., Hugall, J. T., Baumberg, J. J. and Scherman, O. A. (2010). Raman and SERS spectroscopy of cucurbit[*n*]urils, *Phys. Chem. Chem. Phys.*, **12**, pp. 10429–10433.
8. Fierro-Mercado, P. M. and Hernandez-Rivera, S. P. (2012). Highly sensitive filter paper substrate for SERS trace explosives detection, *Int. J. Spectrosc.*, **2012**, 7.
9. Kasera, S., Herrmann, L. O., del Barrio, J., Baumberg J. J. and Scherman, O. A. (2014). Quantitative multiplexing with nano-self-assemblies in SERS, *Sci. Rep.*, **4**, 6785.
10. Cecchini, M. P., Turek, V. A., Paget, J., Kornyshev, A. A. and Edel, J. B. (2013). Self-assembled nanoparticle arrays for multiphase trace analyte detection, *Nat. Mater.*, **12**, pp. 165–171.
11. Peveler, W. J., Jia, H., Jeen, T., Rees, K., Macdonald, T. J., Xia, Z., Chio, W.-I. K., Moorthy, S., Parkin, I. P., Carmalt, C. J., Algar, W. R. and Lee, T.-C. (2019). Cucurbituril-mediated quantum dot aggregates formed by aqueous self-assembly for sensing applications, *Chem. Commun.*, **55**, pp. 5495–5498.

12. Taylor, R. W., Lee, T.-C., Scherman, O. A., Esteban, R., Aizpurua, J., Huang, F. M., Baumberg, J. J. and Mahajan, S. (2011). Precise subnanometer plasmonic junctions for SERS within gold nanoparticle assemblies using cucurbit[n]uril "glue", *ACS Nano*, **5**, pp. 3878–3887.

13. Lin, M. Y., Lindsay, H. M., Weiz, D. A., Ball, R. C., Klein R. and Meakin, P. (1989). Universality in colloid aggregation, *Nature*, **339**, pp. 360–362.

14. Li, X.-B., Li, Z.-J., Gao, Y.-U., Meng, Q.-Y., Yu, S., Weiss, R. G., Tung, C.-H. and Wu, L.-Z. (2014). Mechanistic insights into the interface-directed transformation of thiols into disulfides and molecular hydrogen by visible-light irradiation of quantum dots, *Angew. Chem. Int. Ed. Engl.*, **53**, pp. 2085–2089.

15. Peveler, W. J., Roldan, A., Hollingsworth, N., Porter, M. J. and Parkin, I. P. (2016). Multichannel detection and differentiation of explosives with a quantum dot array, *ACS Nano*, **10**, pp. 1139–1146.

Chapter 6

Case Study: Urinary Biomarkers

This chapter is intended to provide a full case study on macrocycle-functionalised nanosensors based on cucurbiturils (CB7) and gold nanoparticles (Au NPs) for the detection of urinary biomarkers, creatinine (CRN) and uric acid (UA), in water and synthetic urine, via surface-enhanced Raman scattering (SERS) and UV-Vis spectroscopy.

6.1 Introduction

Metabolic information and renal function of the human body can be extracted from our bodily fluids such as urine and blood serum. Compared to blood serum, urine can be obtained in a non-invasive manner. Regular evaluation of different biomolecules may help to diagnose life-threatening diseases and improve the successful rate of medical treatment. Creatinine (CRN), which is the end product of muscle metabolism, is a critical biomarker for monitoring kidney filtration function in renal clearance tests. It is used as the internal standard to normalise the variations of water and other biomarkers in urine owing to its approximately constant excretion rate. Apart from renal function, elevated CRN level is also related to other diseases including muscular dystrophy, hyperthyroidism, and poliomyelitis [1–3]. The current gold standard for CRN sensing in

Macrocycle-Functionalised Nanosensors
Weng-I Katherine Chio and Tung-Chun Lee

ISBN 978-981-5129-33-5 (Hardcover), 978-1-003-51057-4 (eBook)
www.jennystanford.com

clinics and hospitals is based on the Jaffe reaction, a colorimetric technique involving the formation of reddish-orange complexes between CRN and picric acid in an alkaline solution. Methods such as enzymatic technique, capillary electrophoresis, high-performance liquid chromatography (HPLC), and isotope dilution mass spectrometry are also used in clinical studies. However, the most popular Jaffe method is non-specific and subjected to interferences from background molecules such as glucose and proteins, while other methods do not allow highly sensitive and selective point-of-care diagnosis in real time. Uric acid (UA), the end product of metabolism of purine nucleotides, is another important biomarker for renal diseases and other health conditions such as gout, preeclampsia, and hypertension. Commonly used methods for the detection of uric acid include enzymatic technique, HPLC, and capillary electrophoresis, which have drawbacks such as high cost, time-consuming, and requiring sophisticated sample preparation. Therefore, it is of great significance to develop high performance sensors for diagnostic and clinical applications.

6.2 CB7-Au NP Nanosensors for Urinary Biomarkers

Macrocycle-functionalised nanosensors based on plasmonic NPs offer numerous advantages such as high sensitivity, rapid response, cost-effectiveness, minimal sample preparation, and a wider information window for the quantification of biomolecules such as CRN and UA in clinical settings. Macrocycles can overcome the issues of poor sensitivity and reproducibility due to uncontrolled aggregation of NPs by acting as glue to regulate the spacing between NPs while localising biomolecules at the centre of the plasmonic hotspots through host–guest complexation. This section will discuss a case on sensing of urinary biomarkers with nanoaggregates of cucurbit[7]urils (CB7) and gold nanoparticles (Au NPs), covering from formation of host–guest complexes, through macrocycle-mediated nanoaggregation, to UV-Vis and SERS sensing in water and diluted synthetic urine (SU) of clinically relevant concentrations (Fig. 6.1) [4, 5].

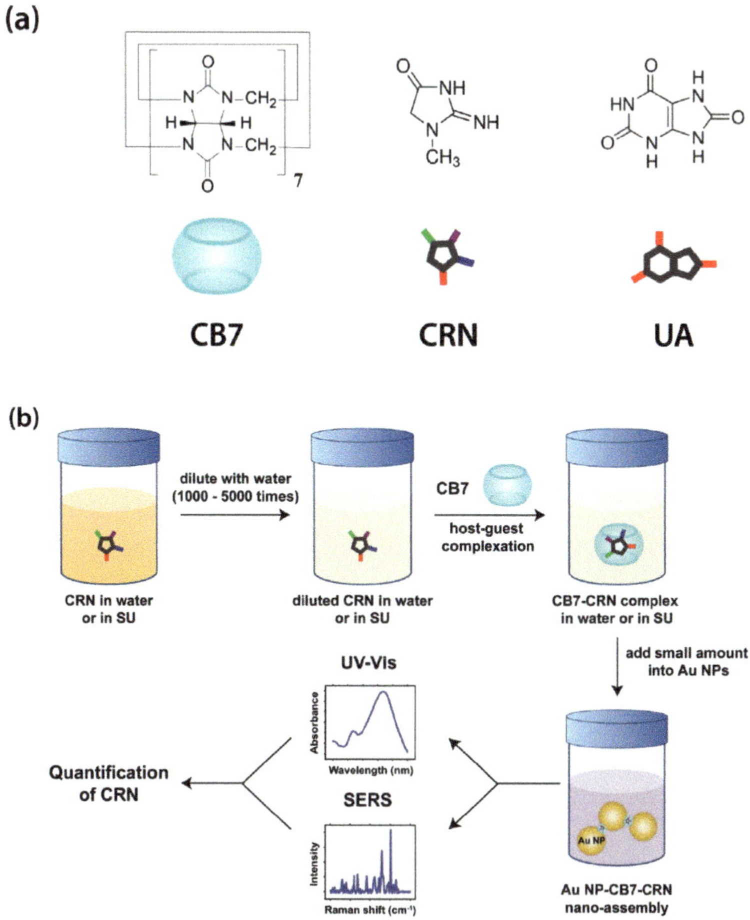

Figure 6.1 Schematic illustrations of (a) CB7, CRN, and UA and (b) the dual UV-Vis and SERS detection of CRN in water or SU using the CB7-Au NP nanosensor.

6.2.1 Supramolecular Chemistry of CB7 and CRN

The host–guest complexation between CB7 and CRN can be verified by the characteristic upfield shifts of the CRN proton signals (H_a and H_b) in the ^1H NMR spectra after mixing them in D_2O with 1:1 stoichiometry (Fig. 6.2). After host–guest complexation, the change in chemical shift ($\Delta\delta$) for H_b (–0.15 ppm) of CRN is much greater than that for H_a (−0.02 ppm), indicating that H_b of the methyl group is deep inside the CB7 cavity, whereas H_a of the CH_2 group is closer to the carbonyl portals. The formation of host–guest complexes in aqueous solution is mainly driven by the release of high-energy water from the CB7 cavity, together with ion-dipole interactions and hydrogen bonding between the electron-rich carbonyl portal of CB7 and the protonated site on CRN (see Section 2.4) [6]. The binding constant of the CB7-CRN complex can be obtained by UV-Vis titration of 50 μM CRN with up to 2 equivalents of CB7, assuming a 1:1 binding model (Fig. 6.3). The relatively low binding constant $(1.17 \pm 0.03) \times 10^4\ M^{-1}$ is owing to the high water solubility and small molecular size of CRN.

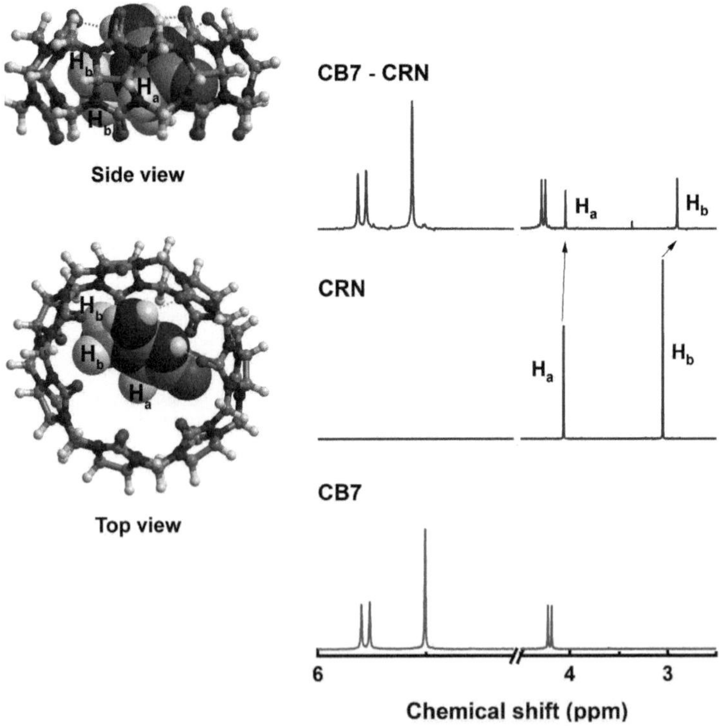

Figure 6.2 DFT molecular model of a $[CB7\text{-}CRN\text{-}H]^+$ host–guest complex in water (left) and ^1H NMR spectra of CB7, CRN, and 1:1 CB7-CRN host–guest complex in D_2O (right).

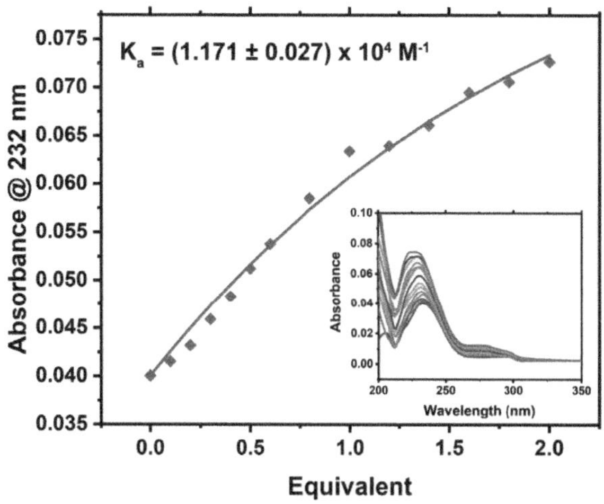

Figure 6.3 UV-Vis titration of CB7 with 50 μM CRN. Inset: UV-Vis spectra of CB7-CRN.

Note that CRN exists in its protonated form in water and SU (pH = 5–7) as the conjugated acid of CRN, $[CRN\text{-}H]^+$, has a pKa of 9.2 [7]. The host–guest binding geometry derived from NMR is supported by the energy-minimised molecular models of the complex in water (CPCM/wB97XD/6-31G*) with a binding energy of –36.12 kcal mol^{-1} (Fig. 6.2). CRN is fully encapsulated within the CB7 cavity, leaving both carbonyl portals available for binding to Au NPs, thus localising CRN at the centre of the plasmonic hotspots which is important for SERS sensing.

6.2.2 Colorimetric Sensing of CB7-CRN Host–Guest Complexes

CRN of clinically relevant concentrations from normal to excess levels (0–300 mg dL^{-1}) in water and SU can be quantified using the CB7-Au NP sensor via UV-Vis spectroscopy (Fig. 6.4, see Table 6.1 for constituents of SU). A constant CB7 concentration of 10 μM is typically used to mediate the formation of reproducible nanoaggregates, thus ensuring signal reproducibility. This phenomenon can be verified by the colour change in the 40 nm citrate-capped Au NP solution from red to purple, and the appearance of a new localised surface

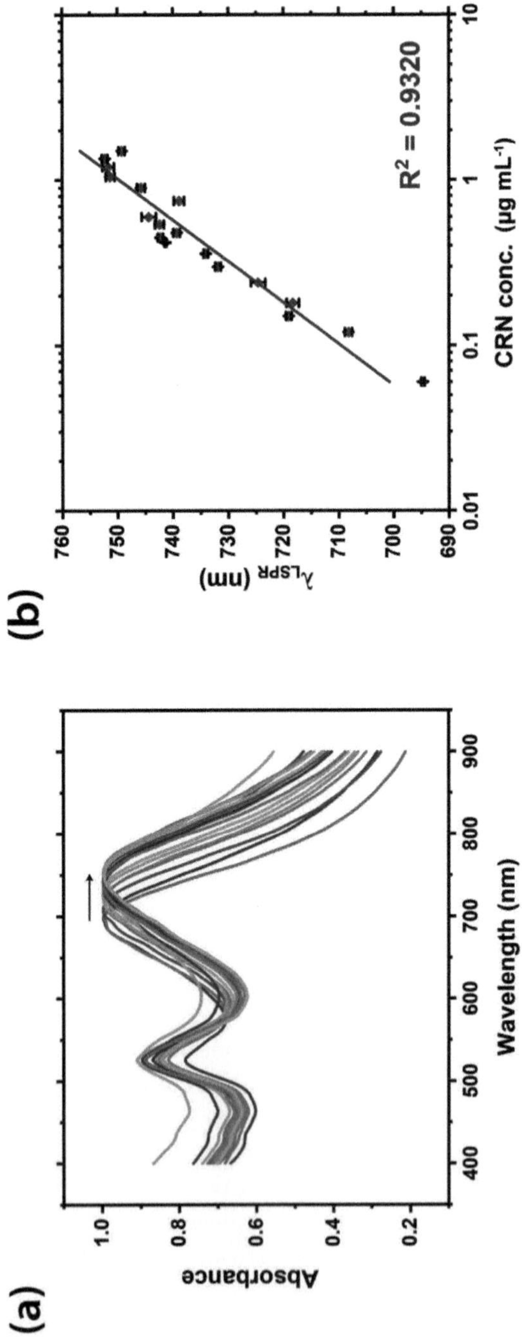
(a)
Absorbance
1.0
0.8
0.6
0.4
0.2
400
500
600
700
800
900
Wavelength (nm)
(b)
λ_{LSPR} (nm)
760
750
740
730
720
710
700
690
0.01
0.1
1
10
CRN conc. (μg mL^{-1})
$R^2 = 0.9320$

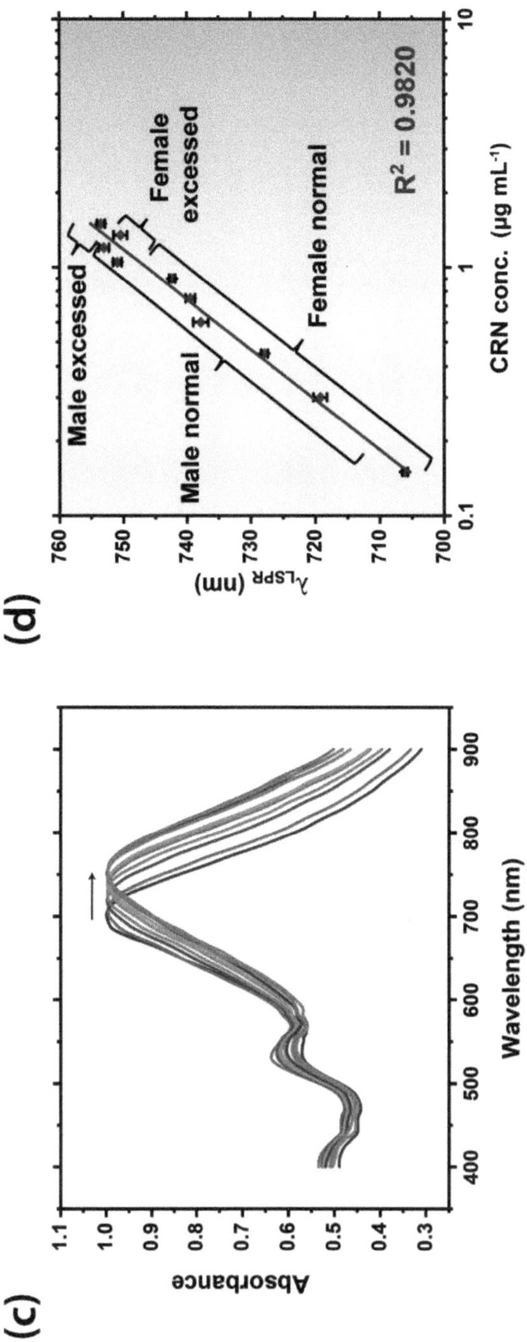
(c)
Absorbance
Wavelength (nm)
(d)
λ_{LSPR} (nm)
CRN conc. (μg mL⁻¹)
Male excessed
Male normal
Female excessed
Female normal
R^2 = 0.9820

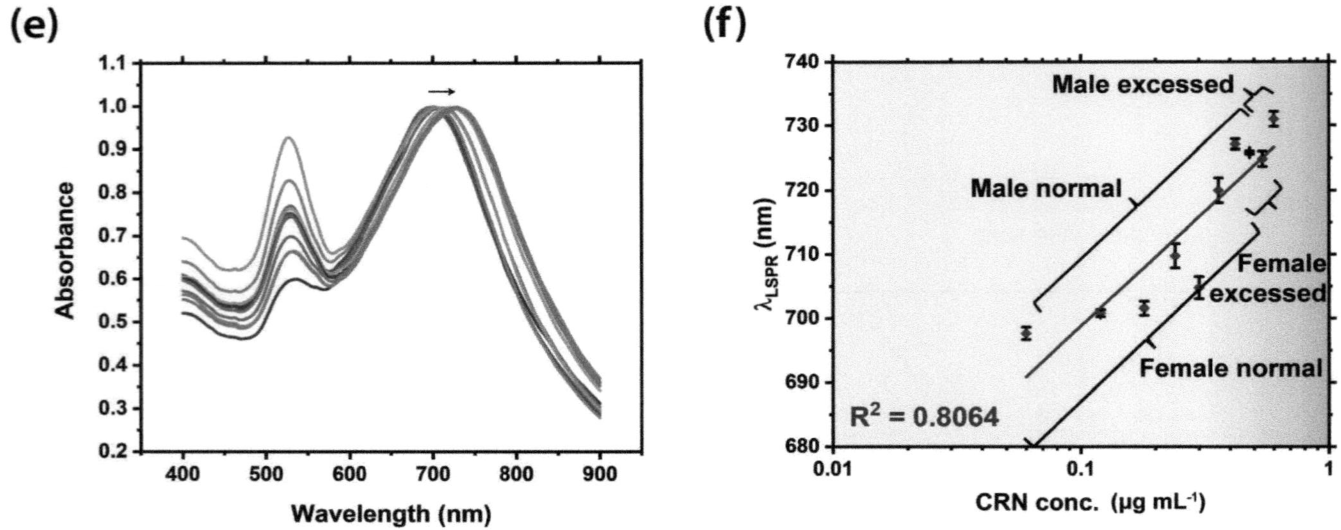

Figure 6.4 (a, c, e) Colorimetric sensing of CRN with concentrations from 0 to 1.50 μg mL^{-1} in (a) water, (c) 2000-fold diluted SU, and (e) 5000-fold diluted SU via UV-Vis spectroscopy, using the CB7-Au NP nanosensor. LSPR peak shifts are indicated by arrows. (b, d, f) Corresponding plots of λ_{LSPR} in (a, c, e) against CRN concentration.

plasmon resonance (LSPR) peak is attributed to the longitudinal plasmonic resonance in the UV-Vis spectra, suggesting the presence of chain-like nanoaggregates with CB7 being doubly-capped by the Au NPs. Logarithmic fittings can be performed to reveal a correlation between the LSPR peak wavelength (λ_{LSPR}) and the concentration of CRN. Notably, there are strong correlations between λ_{LSPR} and log concentration of CRN in water ($R^2 \sim 0.93$) and 2000-fold diluted SU ($R^2 \sim 0.98$), and reasonable correlation in 5000-fold diluted SU ($R^2 \sim 0.81$) respectively (Fig. 6.4). The low detection limit of 0.06 μg mL^{-1} (0.53 μM) with the CB7-Au NP nanosensor allows quantification of CRN in highly diluted SU while minimising the matrix effects, i.e., signal perturbation from other biomolecules and proteins present in the sample (see Section 6.2.6). It should be noted that 1 μg mL^{-1} in measurement samples of 2000-fold diluted SU corresponds to 2000 μg mL^{-1} of the actual CRN concentration in undiluted clinical samples.

Interestingly, CRN alone (>0.30 μg mL^{-1}) can trigger the aggregation of Au NPs without CB7 in water despite the less robust correlation between λ_{LSPR} and log concentration of CRN (Fig. 6.5a). However, it fails to aggregate Au NPs in SU due to preferential surface crowding by other species in the complex matrix (Fig. 6.5b), confirming the critical role of the macrocycle in the formation of nanoaggregates [8]. Attachment of the $[CB7\text{-}CRN\text{-}H]^+$ complexes onto the citrate-capped Au NPs can speed up the aggregation by neutralising the negative surface charges and decreasing the long-range electrostatic repulsion between Au NPs, thus reducing the colloidal stability. This can be verified by the increase in the hydrodynamic size of the nanoaggregates and the decrease in the magnitude of the zeta potential in the presence of CB7 (Fig. 6.6). In addition, $[CRN\text{-}H]^+$ alone can trigger aggregation of Au NPs in a similar mechanism, upon electrostatic interactions or chemisorption with the surface of Au NPs via its nitrogen containing groups.

While colorimetric sensing of CRN via UV-Vis spectroscopy is potentially useful for disease diagnosis, there is no direct evidence to show whether the shift in λ_{LSPR} is caused by CRN or other molecules present in the matrix. For instance, the λ_{LSPR} of 0.60 μg mL^{-1} CRN in 2000-fold and 5000-fold diluted SU are at different wavelengths (738 nm and 731 nm respectively). Nonetheless, this sensing scheme can be used as a quick test to complement and cross-validate the results from the CB7-Au NP SERS nanosensor (see Section 6.2.3).

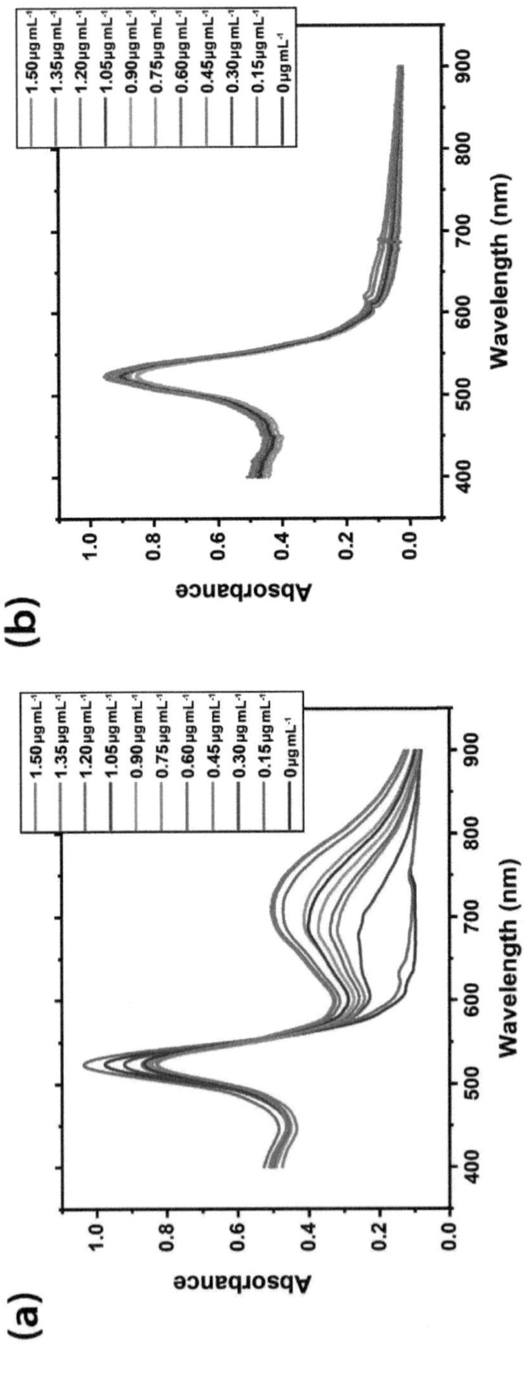

Figure 6.5 UV-Vis spectra of Au NP-CRN in (a) water and (b) 2000-fold diluted SU with CRN concentrations from 0 to 1.50 μg mL^{-1}.

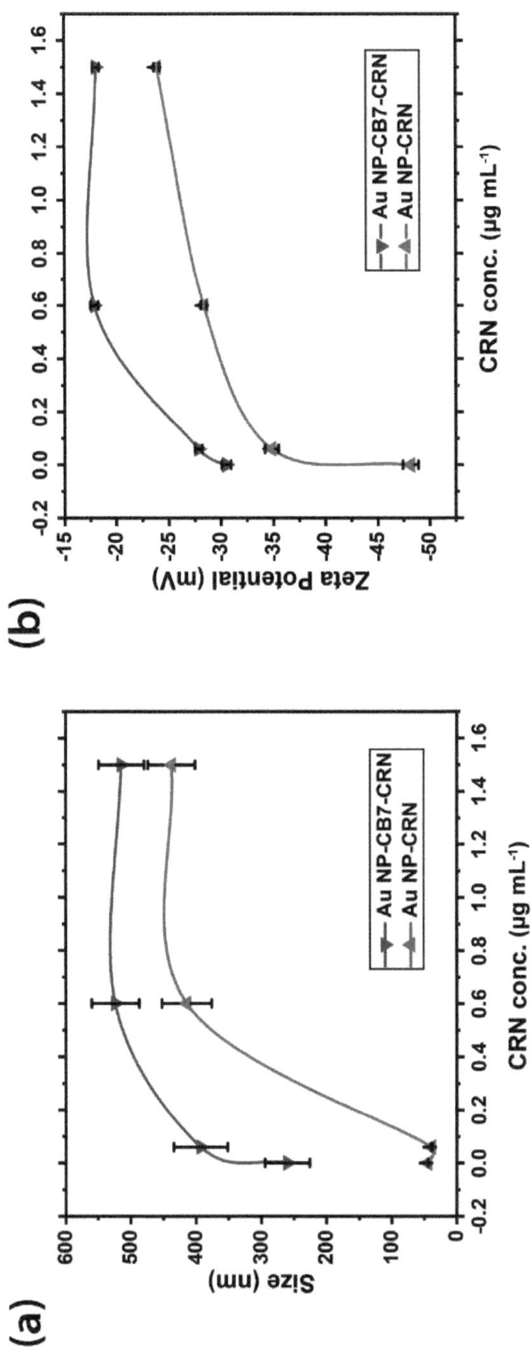

Figure 6.6 (a) Hydrodynamic size and (b) zeta potential of Au NP-CRN and CB7-Au NP-CRN with CRN concentrations from 0 to 1.50 μg mL^{-1}.

Table 6.1 Components dissolved in 90 mL of water to prepare 100 mL SU

Component	Mass (mg)	Component	Mass (mg)
Urea	900	Citric acid	40
Sodium chloride	500	Uric acid	16
Sodium phosphate	480	Albumin	5
Potassium chloride	450	Creatinine	0–300

6.2.3 SERS Sensing of CB7-CRN Complexes in Water

In contrast to colorimetric techniques, SERS sensing is highly specific as it allows identification of the target analytes (i.e., CRN) via their vibrational fingerprints, thus differentiating them from other biomarkers (e.g., UA) in the sample matrix. CRN powders can be characterised by multiple peaks in the Raman spectrum at 608, 674, 839, 904, 1246, 1416, 1478, 1657, and 1712 cm^{-1} (Fig. 6.7a). CB7 crystals are characterised by two major peaks at 444 and 833 cm^{-1}, corresponding to the ring scissor and ring deformation modes (Fig. 6.7b). Note that it is often challenging to measure the Raman signals of CB7, CRN, and CB7-CRN complexes in aqueous media owing to the intrinsically low Raman scattering cross sections (see Section 4.6).

Instead, SERS signals can be measured by mixing CB7, CRN, or pre-formed 1:1 CB7-CRN solution with 40 nm citrate-capped Au NP solution. CB7 defines precise nanojunctions between the Au NPs by capping them with the electron-rich carbonyl portals, leading to very strong and localised plasmonic hotspots within the nanoaggregates. Compared to the Raman peaks of CB7 crystals and CRN powders, the SERS peaks of the respective solutions are broadened and shift slightly, due to the solution effect and the molecular interaction between CB7 and Au NPs (Fig. 6.8a) [9]. The total SERS signals of CB7 and CRN are approximately equal to those of the CB7-CRN complexes, implying that the major tautomer of CRN (i.e., amino tautomer, 2-amino-1-methyl-2-imidazoline-4-one) [10] remains the same upon binding to CB7 in aqueous solution. In addition, the Raman spectra of CB7, CRN, and the CB7-CRN complexes modelled in water at CPCM/wB97XD/6-31G* level of theory show general consistency with the experimental results (Fig. 6.8b).

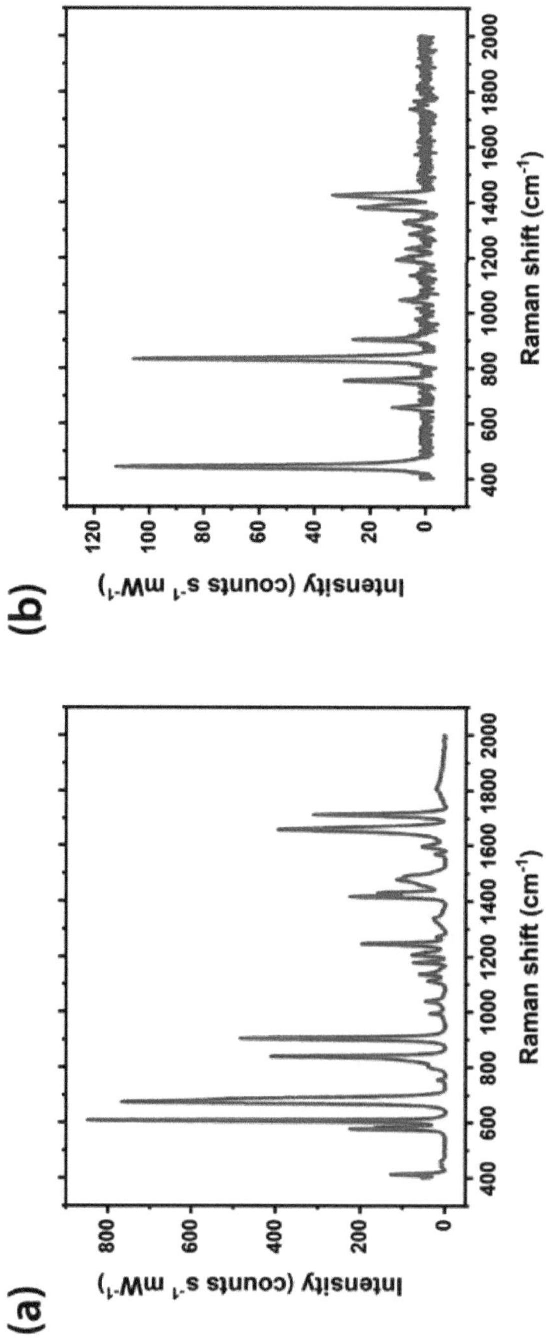

Figure 6.7 Raman spectra of (a) CRN powders and (b) CB7 crystals.

(a)

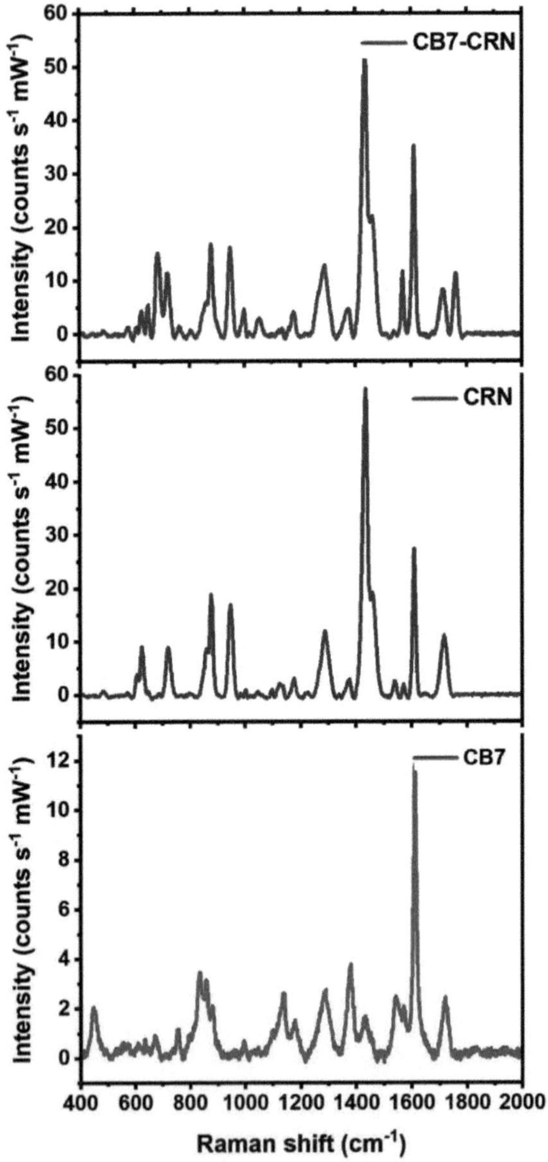
CB7-CRN
CRN
CB7
Intensity (counts s⁻¹ mW⁻¹)
Raman shift (cm⁻¹)
0
10
20
30
40
50
60
2
4
6
8
12
400
600
800
1000
1200
1400
1600
1800
2000

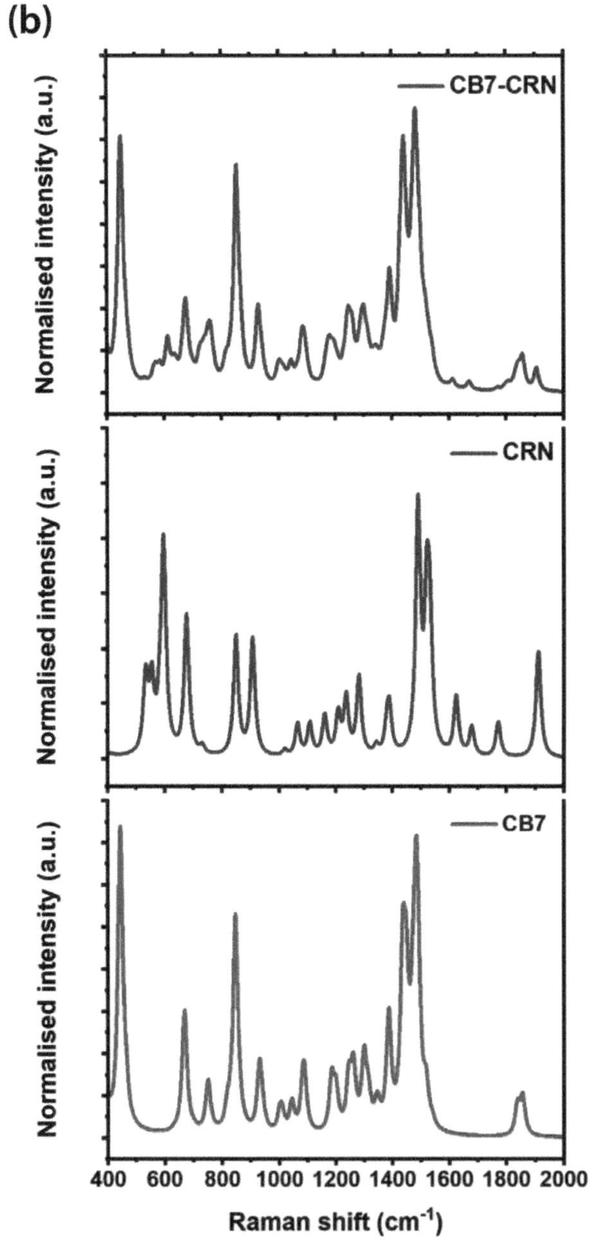

Figure 6.8 (a) Experimental SERS spectra of CB7, CRN, and 1:1 CB7-CRN host–guest complexes in 40 nm Au NP solution. (b) Calculated Raman spectra of CB7, CRN, and 1:1 CB7-CRN host–guest complexes in water at CPCM/wB97XD/6-31G* level of theory.

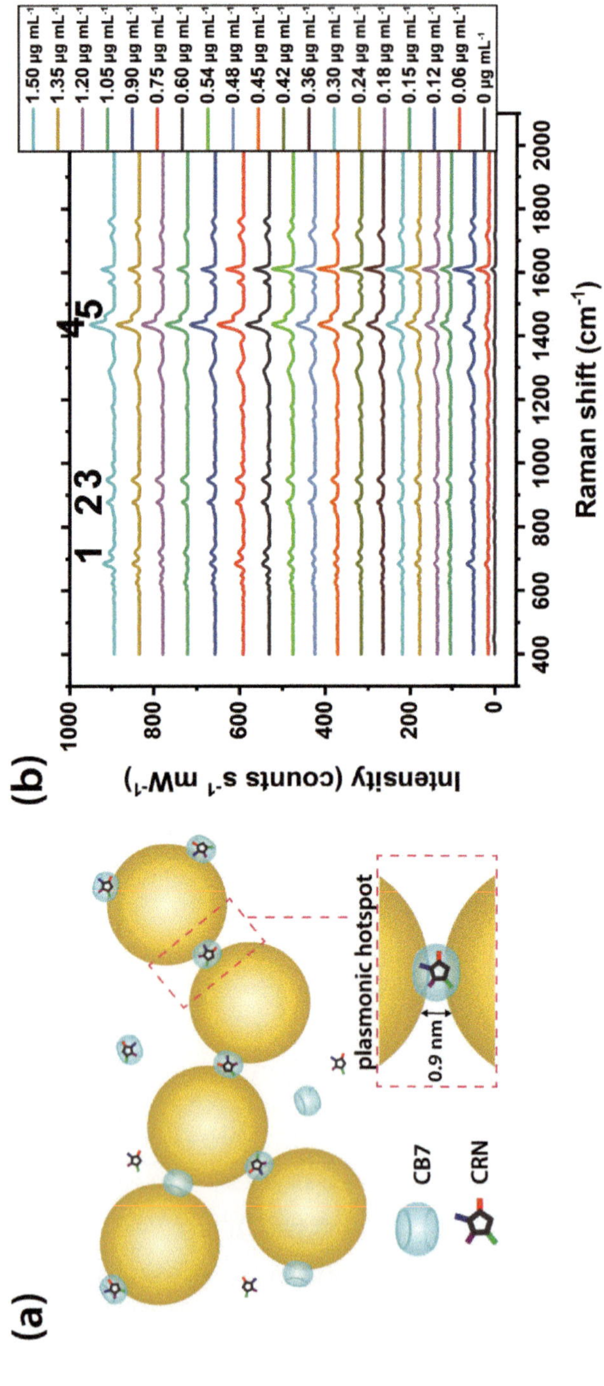
(a)
CB7
CRN
plasmonic hotspot
0.9 nm
(b)
Intensity (counts s-1 mW-1)
Raman shift (cm-1)
1
2 3
4
5
1.50 μg mL-1
1.35 μg mL-1
1.20 μg mL-1
1.05 μg mL-1
0.90 μg mL-1
0.75 μg mL-1
0.60 μg mL-1
0.54 μg mL-1
0.48 μg mL-1
0.45 μg mL-1
0.42 μg mL-1
0.36 μg mL-1
0.30 μg mL-1
0.24 μg mL-1
0.18 μg mL-1
0.15 μg mL-1
0.12 μg mL-1
0.06 μg mL-1
0 μg mL-1

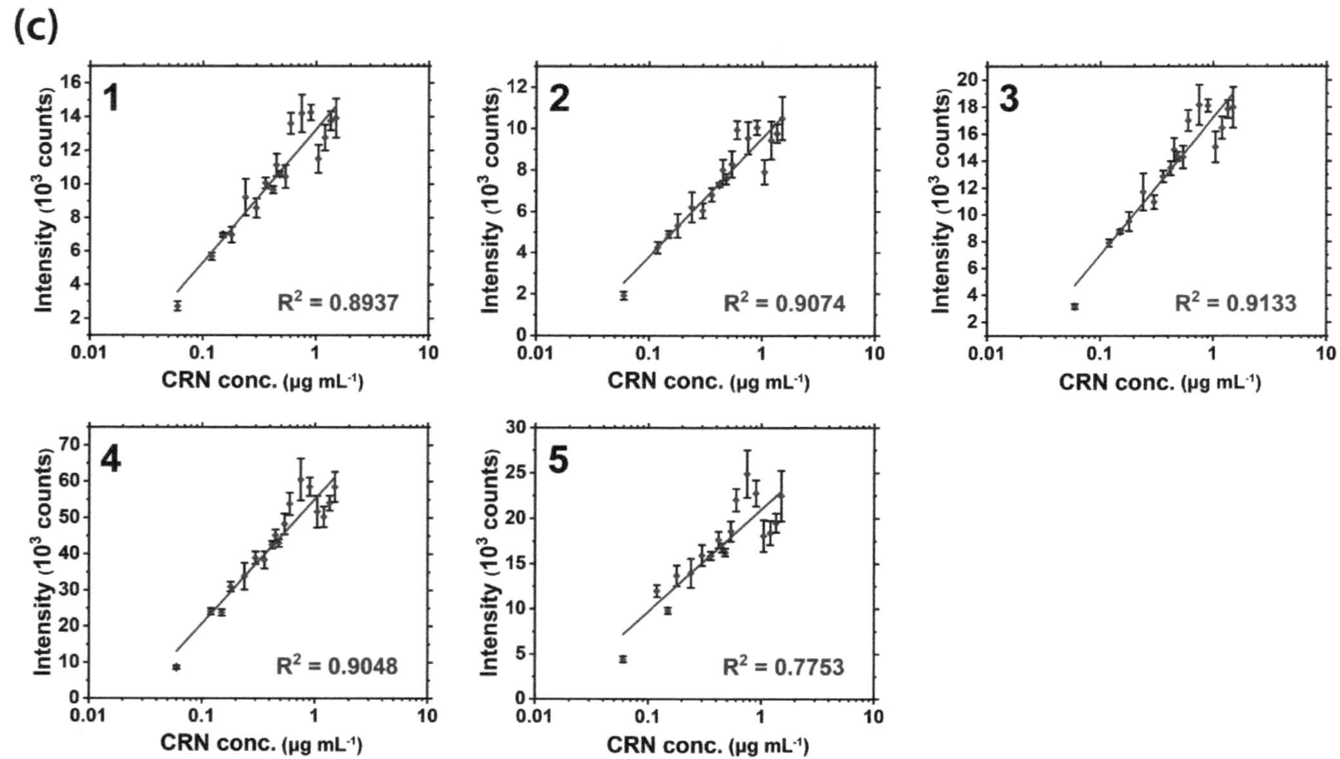

Figure 6.9 (a) Schematic illustration of the precise plasmonic hotspots within the CB7-Au NP nanosensor for CRN detection in water (not to scale). (b) SERS spectra of CRN with concentrations from 0 to 1.50 μg mL^{-1} (13.3 μM) in water. (c) Corresponding plots of SERS intensity of five characteristic CRN peaks in (b) against its concentration.

The sensitivity of the CB7-Au NP nanosensors for biomarkers in water can be demonstrated by performing SERS measurements of CRN with concentrations from 0 to 1.50 μg mL^{-1} (13.3 μM) at 10 μM CB7 (Fig. 6.9). A constant concentration of the macrocycle is needed to ensure the formation of reproducible SERS substrates. Although most of the CRN molecules are fully encapsulated within the CB7 cavity, there might be a small amount of free (i.e., uncomplexed) CRN trapped within the CB7-Au NP nanoaggregates when CRN is in excess. The major peaks of CRN can be clearly observed even when the concentration is down to 0.06 μg mL^{-1} (0.53 μM) with a good signal-to-noise ratio (SNR) of 12.8, implying that the actual detection limit of the CB7-Au NP nanosensors for CRN could be down to 4.69 ng mL^{-1} (41.5 nM). The SNR is defined as the ratio of the intensity of the major CRN peak at 1434 cm^{-1} to that of the background signals at 1800–2000 cm^{-1}. Logarithmic fittings are usually performed to reveal a correlation between SERS intensity and concentration of CRN. Note that there are good correlations ($R^2 \sim 0.78$–0.91) between the SERS intensity and the log concentration of CRN for multiple characteristic peaks at 721, 877, 947, 1435, and 1461 cm^{-1} but a slight drop in signal linearity when the CRN concentration is ≥1.05 μg mL^{-1} (9.28 μM), probably due to shifting of λ_{LSPR} from the laser excitation (Fig. 6.9c) [11].

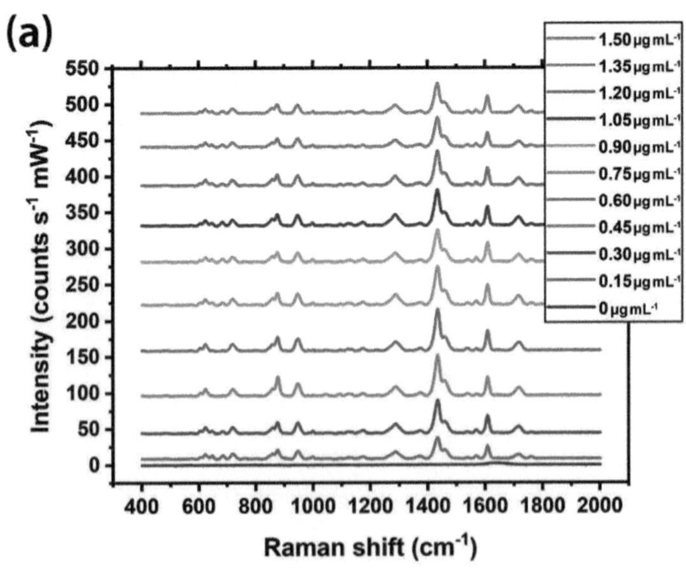

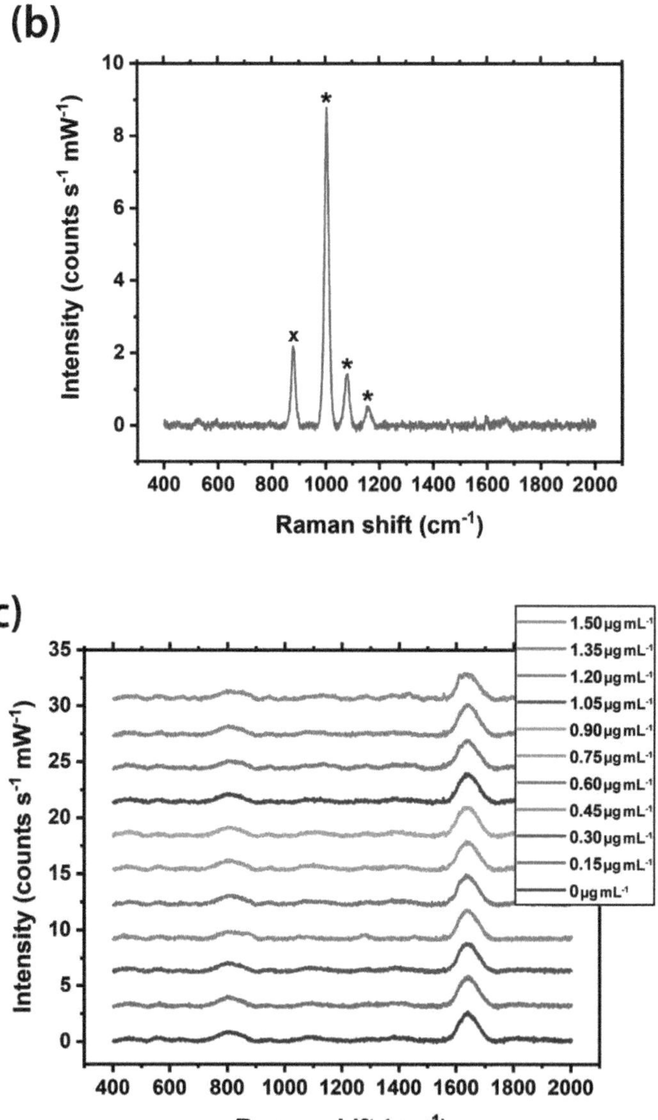

Figure 6.10 (a) SERS sensing of CRN with concentrations from 0 to 1.50 μg mL^{-1} in water in the absence of CB7. (b) Raman spectrum of SU. Main Raman peaks of sodium phosphate and urea are marked by x and *, respectively. (c) SERS sensing of CRN with concentrations from 0 to 1.50 μg mL^{-1} in 2000-fold diluted SU in the absence of CB7.

Moreover, CRN can still be detected in the absence of CB7 as the electrostatic interactions or chemisorption of CRN with the surface of Au NPs causes nanoaggregation, but a correlation between the SERS intensity and concentration of CRN is lacking (Fig. 6.10a).

6.2.4 SERS Sensing of CB7-CRN Complexes in Urinalysis

The CB7-Au NP nanosensors allow quantification of CRN in SU (i.e., in the presence of other biomolecules and proteins), mimicking the samples in clinical and point-of-care applications. Note that urea and phosphate in SU can be detected under normal Raman spectroscopy [12, 13] while the CRN signals are too weak to be detected (Fig. 6.10b). The prepared SU can be highly diluted with water (1000–5000-fold), so only a very small sample volume is required for the highly sensitive SERS nanosensor. This has important implications for the health monitoring of critically ill patients who often fail to excrete a normal volume of urine, along with other benefits such as minimising the matrix effect, reducing the errors from water evaporation, and loss of samples due to liquid transfer.

In contrast to the detection of CRN in water, no SERS signals of CRN can be observed in diluted SU without CB7, owing to preferential surface crowding by other biomolecules and proteins present in the matrix (Fig. 6.10c). Note that CRN can be detected by the CB7-Au NP SERS nanosensors in 1000-, 2000- and 5000-fold diluted SU, where 1 μg mL^{-1} in measurement samples of 1000-, 2000-, and 5000-fold diluted SU corresponds to 1000, 2000, and 5000 μg mL^{-1} of the actual CRN concentration in undiluted human samples, respectively. Logarithmic fittings are often used to reveal a correlation between SERS intensity and concentration of CRN.

In 1000-fold diluted SU, CRN of concentrations from 0 to 3.00 μg mL^{-1} (26.5 μM) can be quantified by the nanosensor with good correlations ($R^2 \sim 0.81$–0.94) between the SERS intensity of the five characteristic peaks and log concentration of CRN (Fig. 6.11). The relatively large error bars for the samples containing high CRN concentrations are due to the presence of excess $[CRN\text{-}H]^+$ with a constant CB7 concentration, leading to faster aggregation of Au NPs and hence a narrower measurement window. In order to avoid the shift in aggregation kinetics caused by the relatively high ratio of

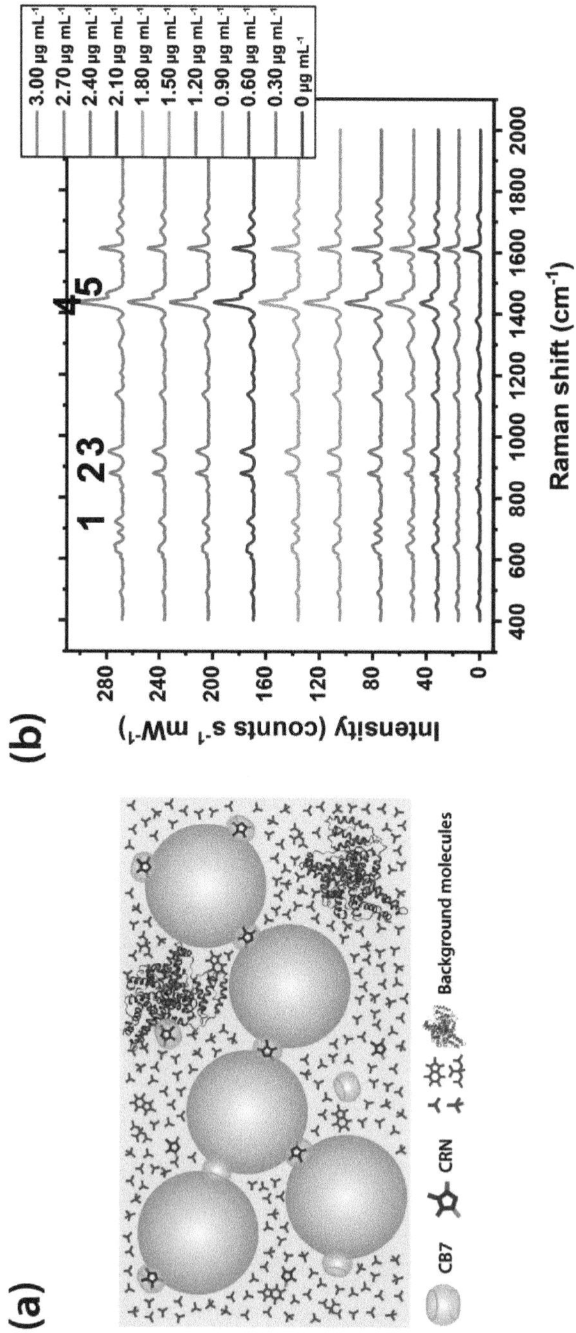
(a)
CB7
CRN
Background molecules
(b)
Intensity (counts s^{-1} mW^{-1})
Raman shift (cm^{-1})
1
2
3
4
5
3.00 μg mL^{-1}
2.70 μg mL^{-1}
2.40 μg mL^{-1}
2.10 μg mL^{-1}
1.80 μg mL^{-1}
1.50 μg mL^{-1}
1.20 μg mL^{-1}
0.90 μg mL^{-1}
0.60 μg mL^{-1}
0.30 μg mL^{-1}
0 μg mL^{-1}

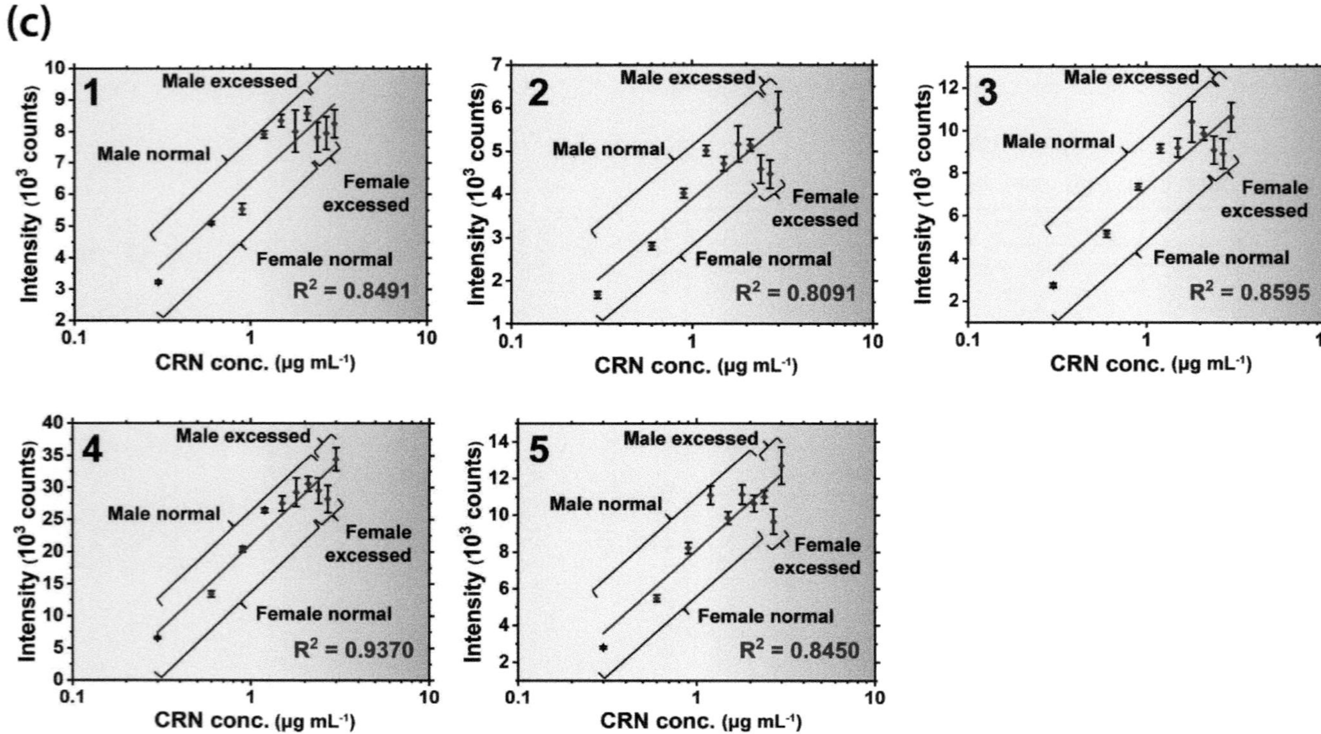

Figure 6.11 (a) Schematic illustration of the precise plasmonic hotspots within the CB7-Au NP nanosensor for CRN detection in 1000-fold diluted SU (not to scale). (b) SERS spectra of CRN with concentrations from 0 to 3.00 μg mL^{-1} (26.5 μM) in 1000-fold diluted SU. (c) Corresponding plots of SERS intensity of five characteristic CRN peaks in (b) against its concentration.

CRN to CB7, CRN sensing is then performed in 2000-fold diluted SU with the concentration of CRN ranging from 0 to 1.50 μg mL^{-1} (13.3 μM). There are strong correlations (R^2 > 0.95) between the SERS intensity and the log concentration of CRN for the characteristic peaks with very small error bars (Fig. 6.12), indicating the high signal reproducibility (<5% error) of the CB7-Au NP nanosensor.

The detection limit of the CB7-Au NP nanosensors for CRN can be tested by performing SERS measurements in 5000-fold diluted SU with CRN concentrations from 0 to 0.60 μg mL^{-1} (5.3 μM). Similarly, strong correlations (R^2 ~ 0.88–0.96) can be observed for all characteristic CRN peaks with small error bars (Fig. 6.13). Note that the major peaks of CRN can be observed in the SERS spectra even when the concentration is down to 0.06 μg mL^{-1} (0.53 μM) with a SNR of 4.8, indicating that the actual detection limit in diluted SU could be as low as 12.5 ng mL^{-1} (111 nM), the lowest among the state-of-the-art methods (Table 6.2).

Table 6.2 Detection limit of CRN in solution using SERS sensors

Type of SERS sensor	Matrix	Detection range (μg mL^{-1})	Detection limit (μg mL^{-1})	Ref.
Au NPs	SU	12.5–50	12.5	[2]
Au NPs	SU	384–1536	384	[14]
	Human urine	25.6–1152†	25.6	
Ag NPs	Water	100–2800	100	[15]
Jaffe complex on Ag film	Water	2.83–16.97	2.83	[16]
Ag-coated	SU	0.5–10.2‡	0.5	[17]
parylene	Human urine	61–1730	61	
Nanoporous	Water	0.011–11.31	0.0015*	[8]
Au disk	SU	1.13–22.62	1.13	
	Mouse urine	20.8–303.0	6.80	
BN-Au nanocomposite	SU	1.13 - 22.62	1.13	[18]
Superhydrophobic Ag film	Water	0.57–1131	0.57	[19]
Au NP-coated	Water	0.20–1.00	0.032¶	[20]
DVD	SU	0.10–10.00‡	0.10	
CB7-Au NPs	Water	0.06–1.50	0.0047*	[4]
	SU	0.06–1.50	0.0125*	

†Linear region: 2.56–6.40 mg dL^{-1}. ‡No linear plot shown. *Calculated from SNR. ¶Calculated from equation.

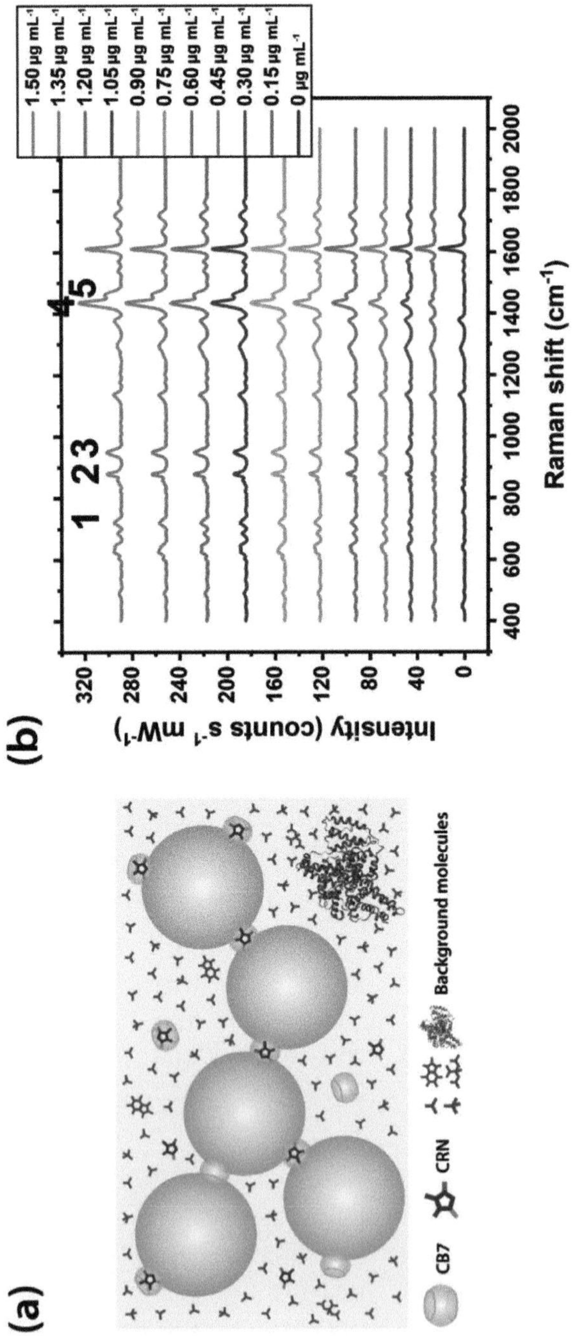
(a)
CB7
CRN
Background molecules
(b)
Intensity (counts s⁻¹ mW⁻¹)
Raman shift (cm⁻¹)
1
2
3
4
5
1.50 μg mL⁻¹
1.35 μg mL⁻¹
1.20 μg mL⁻¹
1.05 μg mL⁻¹
0.90 μg mL⁻¹
0.75 μg mL⁻¹
0.60 μg mL⁻¹
0.45 μg mL⁻¹
0.30 μg mL⁻¹
0.15 μg mL⁻¹
0 μg mL⁻¹

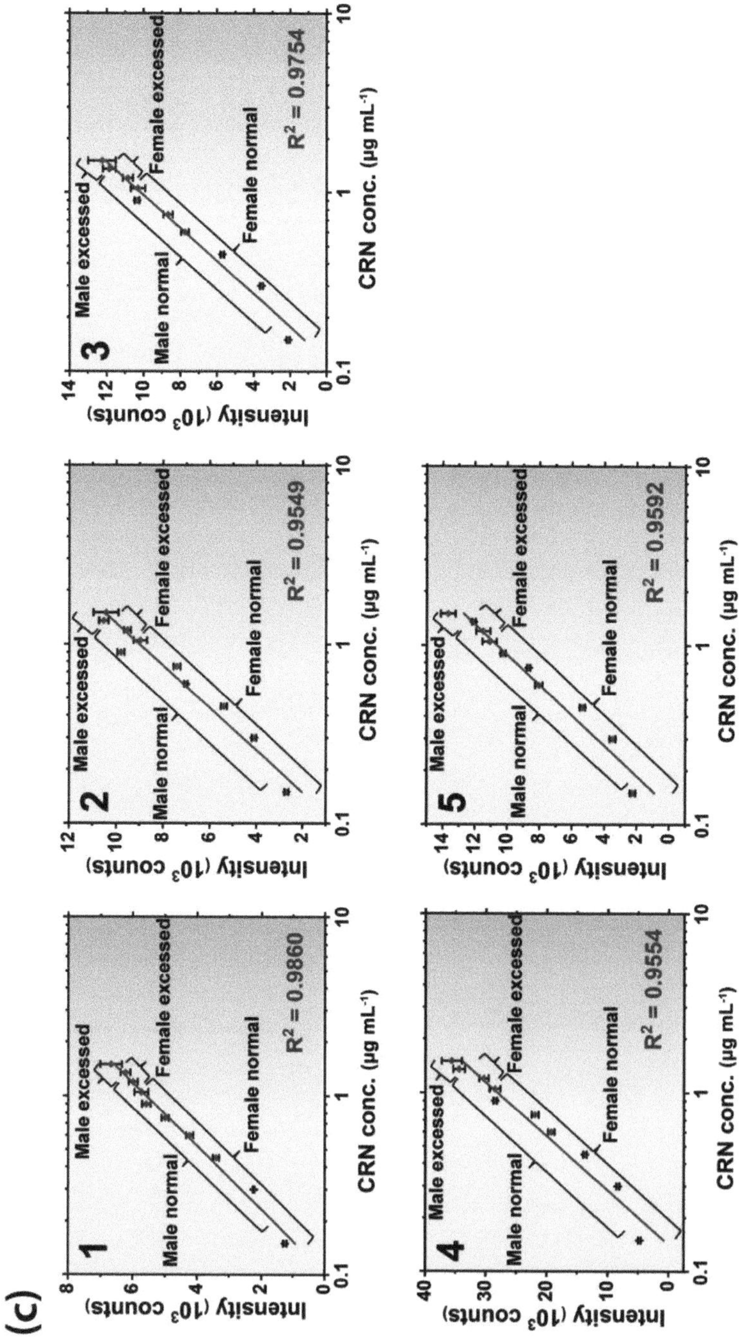

Figure 6.12 (a) Schematic illustration of the precise plasmonic hotspots within the CB7-Au NP nanosensor for CRN detection in 2000-fold diluted SU (not to scale). (b) SERS spectra of CRN with concentrations from 0 to 1.50 μg mL^{-1} (13.3 μM) in 2000-fold diluted SU. (c) Corresponding plots of SERS intensity of five characteristic CRN peaks in (b) against its concentration.

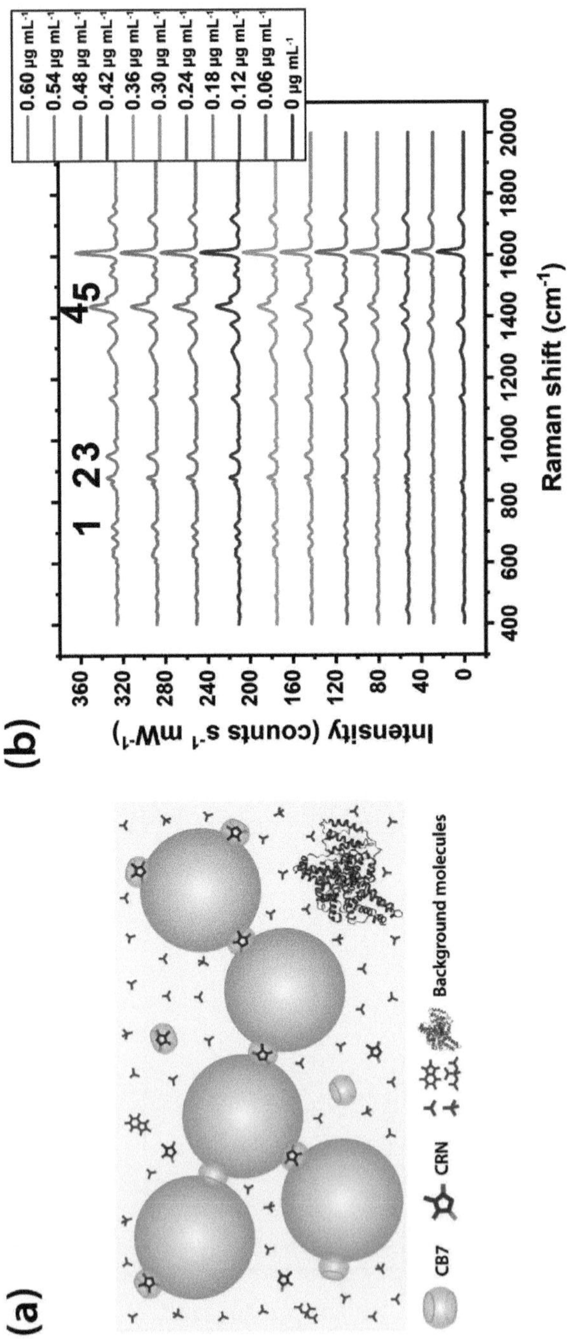
(a)
CB7
CRN
Background molecules
(b)
Intensity (counts s-1 mW-1)
Raman shift (cm-1)
1
2
3
4
5
0.60 μg mL-1
0.54 μg mL-1
0.48 μg mL-1
0.42 μg mL-1
0.36 μg mL-1
0.30 μg mL-1
0.24 μg mL-1
0.18 μg mL-1
0.12 μg mL-1
0.06 μg mL-1
0 μg mL-1

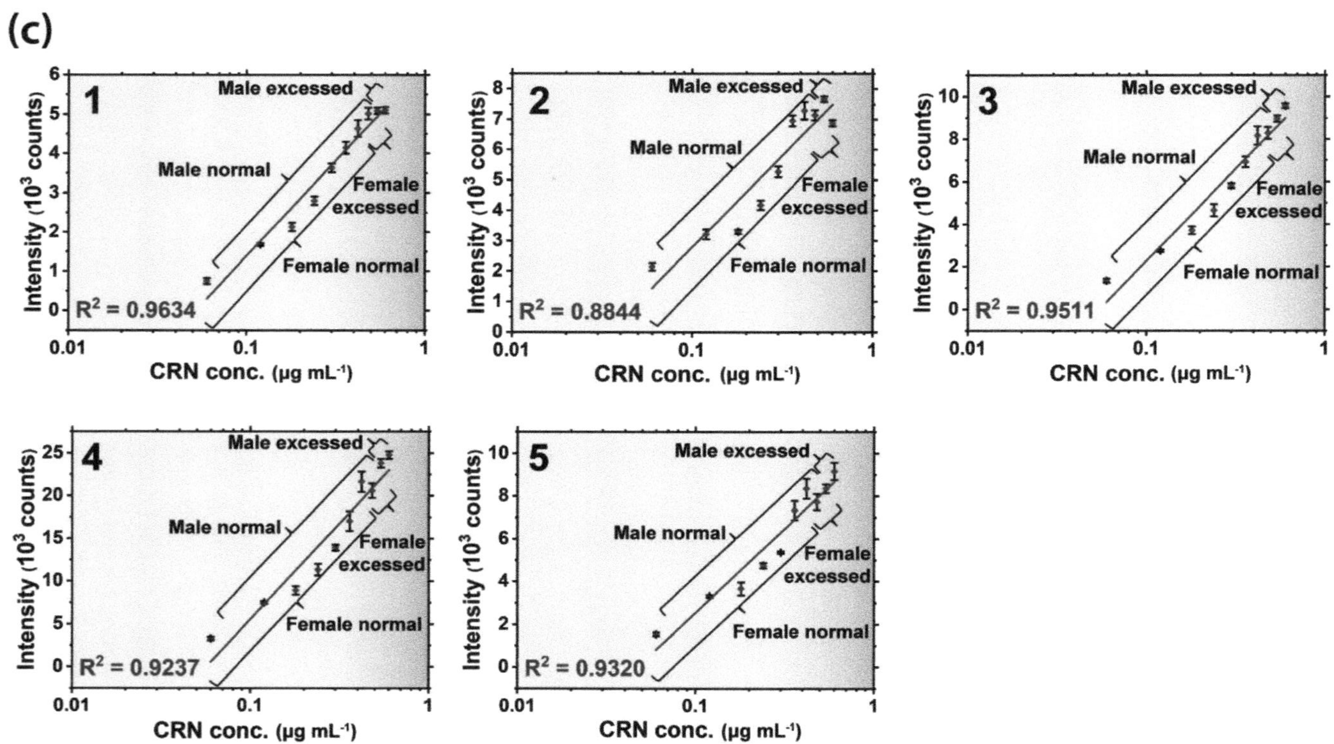

Figure 6.13 (a) Schematic illustration of the precise plasmonic hotspots within the CB7-Au NP nanosensor for CRN detection in 5000-fold diluted SU (not to scale). (b) SERS spectra of CRN with concentrations from 0 to 0.60 µg mL^{-1} (5.3 µM) in 5000-fold diluted SU. (c) Corresponding plots of SERS intensity of five characteristic CRN peaks in (b) against its concentration.

6.2.5 SERS Sensing of CB7-UA Complexes in Water

Similar to the case of CRN, UA can be detected by the CB7-Au NP SERS nanosensor in water via CB7-UA host–guest complexation with empty CB7 mediating the formation of precise plasmonic hotspots (Fig. 6.14a). However, no UA signals can be observed in the absence of the macrocycle because UA alone cannot trigger the aggregation of Au NPs. The SERS measurements are performed at a constant CB7 concentration of 20 μM instead of 10 μM, as the absolute concentration needed is dependent on the analyte molecules and background molecules in the matrix, if any. In particular, a higher concentration of CB7 can be used to speed up the aggregation of Au NPs (see Section 3.4.1). The detection limit of the nanosensor for UA detection is down to ~0.2 μM where the SERS signals of the characteristic UA peaks at 640 cm^{-1} and 1130 cm^{-1}, attributed to skeletal ring deformation and C-N vibration, respectively, can still be observed (Fig. 6.14b–d). The normal concentration of UA present is 16–100 mg dL^{-1} (0.95–5.95 mM) in urine and 3.5–7.0 mg dL^{-1} (0.21–0.42 mM) in blood serum [21, 22], indicating the sensitivity of the nanosensor is good enough for clinical applications. Strong correlations ($R^2 > 0.98$) between the SERS intensity and log concentration of UA can be obtained by power law for both peaks, with linear regions in 0.2–2 μM (Fig. 6.14e,f). Note that the linear correlations can be approximated for a narrow range of concentrations, where the SERS intensity approaches zero with the concentration of analyte molecules (i.e., log concentration tends to negative infinity). The small error bars also indicate that the SERS signals are highly reproducible.

In addition, fundamental properties of the CB7-UA host–guest complexes can be studied through characterisation techniques such as ^{13}C NMR and mass spectrometry, but not ^{1}H NMR due to the lack of non-exchangeable protons in UA. The binding constants can be measured using UV-Vis spectroscopy and isothermal titration calorimetry (ITC). Computational simulations can also be performed to obtain binding geometry and Raman spectra via force-field and DFT modelling if needed.

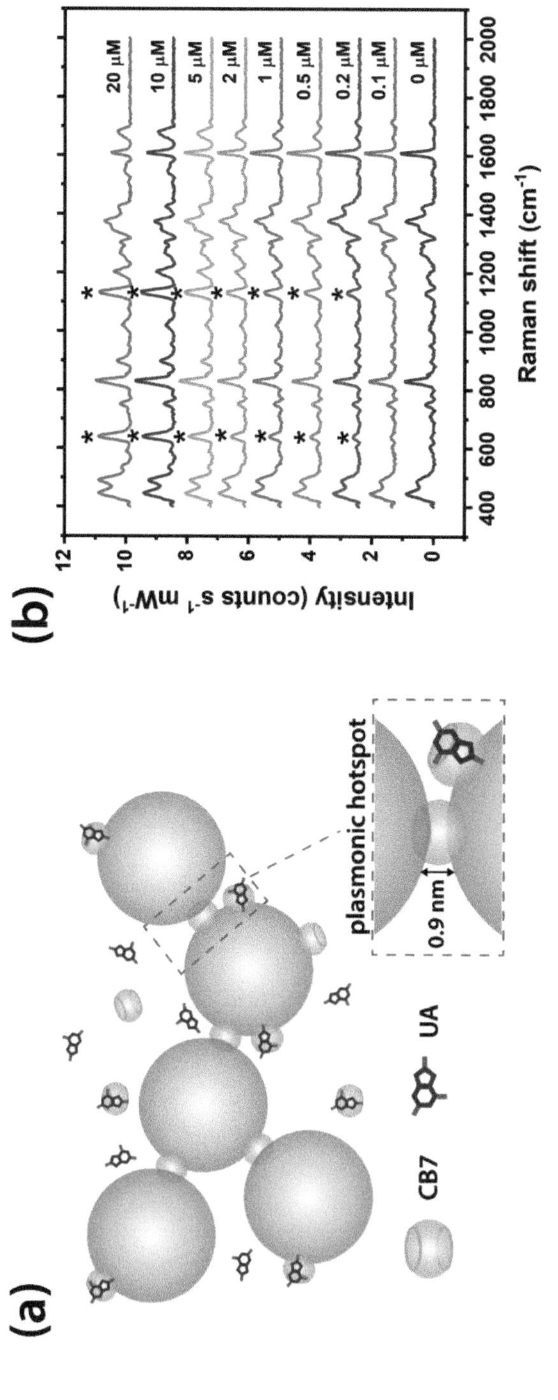
(a)
CB7
UA
plasmonic hotspot
0.9 nm
(b)
Intensity (counts s^{-1} mW^{-1})
Raman shift (cm^{-1})
20 μM
10 μM
5 μM
2 μM
1 μM
0.5 μM
0.2 μM
0.1 μM
0 μM

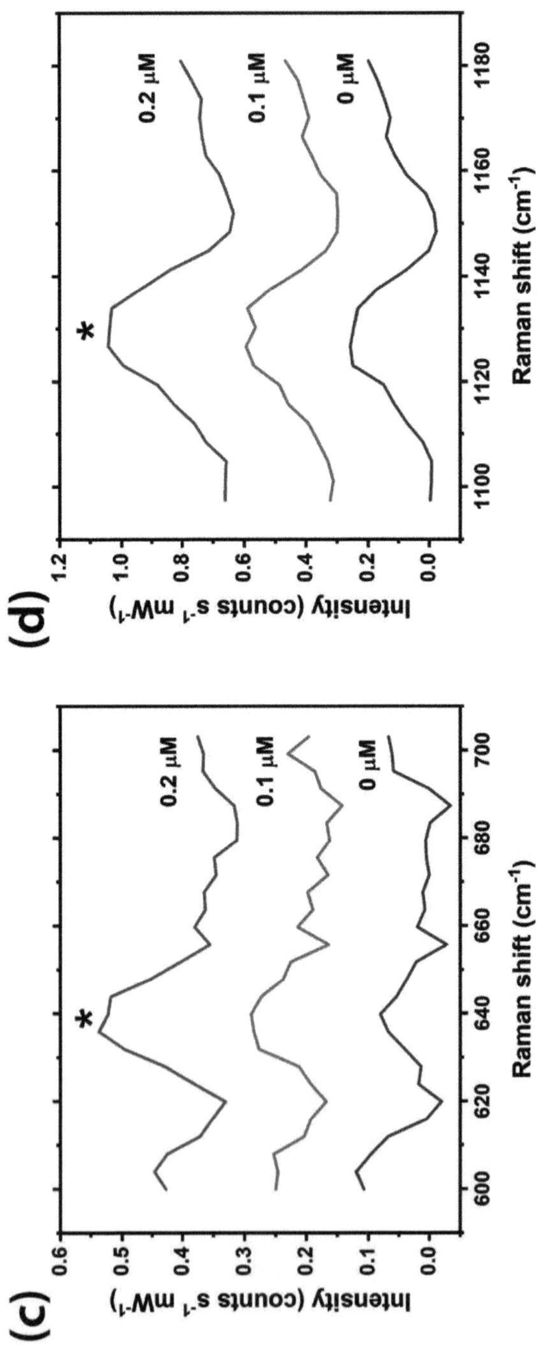
(c)
Intensity (counts s-1 mW-1)
Raman shift (cm-1)
0.2 μM
0.1 μM
0 μM
*
(d)
Intensity (counts s-1 mW-1)
Raman shift (cm-1)
0.2 μM
0.1 μM
0 μM
*

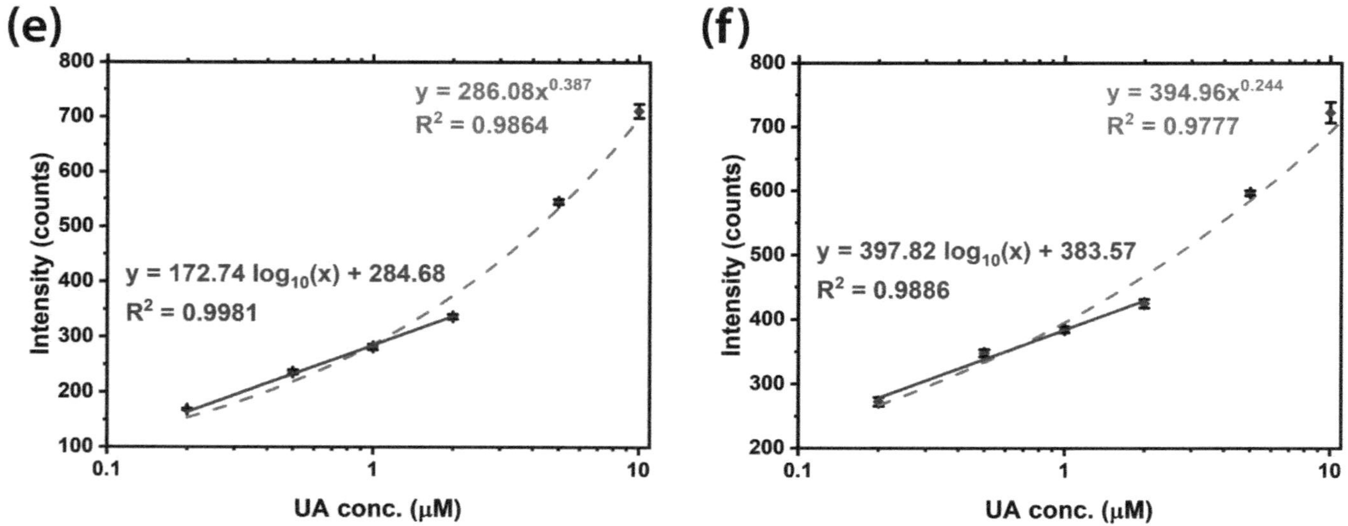

Figure 6.14 (a) SERS spectra of UA with or without CB7. Raman peaks of UA and CB7 are marked by * and +, respectively. (b) Full-range and (c, d) zoom-in SERS spectra of UA with concentrations from 0 to 20 μM. (e, f) Corresponding plots of the SERS intensity of the two main UA peaks in (b) against its concentration.

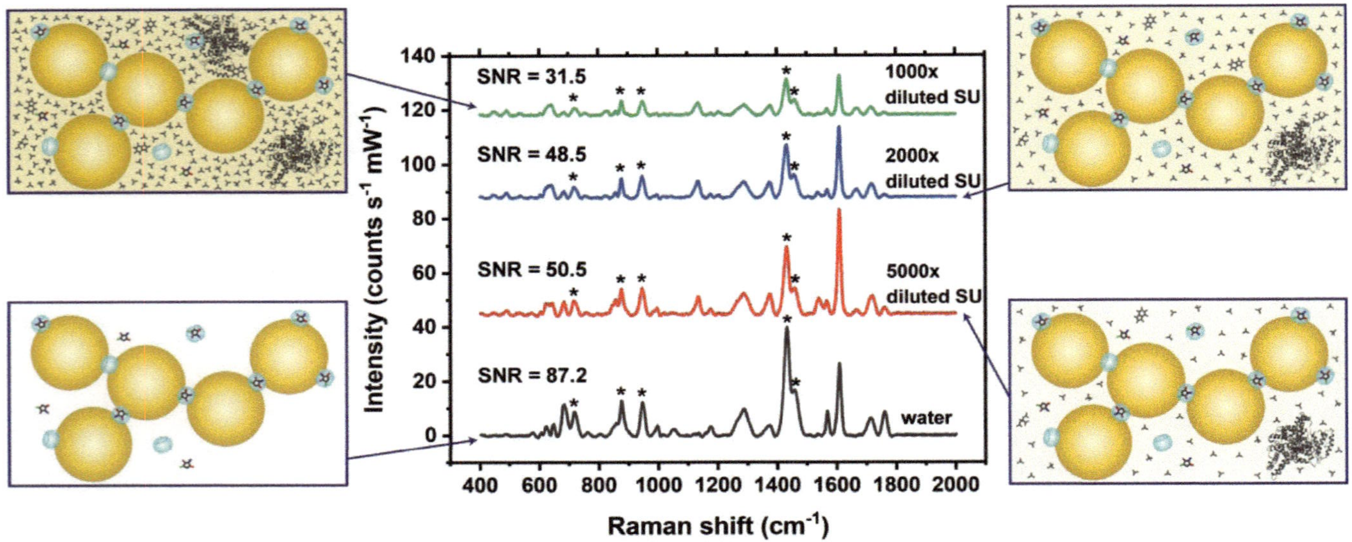

Figure 6.15 SERS sensing of 0.60 μg mL^{-1} CRN in water and sequentially (1000-,2000-, and 5000-fold) diluted SU. Insets: Schematic illustrations of the CB7-Au NP nanosensors for CRN detection in water and sequentially diluted SU, showing the matrix differences (not to scale).

6.2.6 Matrix Effect on SERS Signals

The matrix effect can be verified by comparing the SERS intensity of the CRN signals in water and sequentially diluted SU (Fig. 6.15). Note that the CRN signals with a concentration of 0.60 μg mL^{-1} in 2000-fold diluted SU are 34.0–52.6% lower than those in water, owing to the competitive binding of CRN and other biomolecules to CB7 and Au NPs in the matrix [23]. For example, uric acid can form host–guest complexes with CB7 (see Section 6.2.5) while phosphate may bind to the surface of Au NPs. Since CB7 can bind to both CRN and UA, it is possible to observe interferences in the SERS signals. The matrix effect on CRN sensing can be minimised by increasing the dilution factor, thus recovering the CRN signals in water (without other species) and increasing the SNR. In addition, sample purification such as gel electrophoresis and HPLC can be performed prior to SERS sensing if necessary. Nevertheless, the CB7-Au NP nanosensors can be used for multiplexed detection of biomarkers in a complex matrix such as bodily fluids for clinical applications, especially when coupled to advanced data analysis techniques (see Section 7.2.5).

6.3 Summary

- The CB7-Au NP nanosensor based on UV-Vis spectroscopy and SERS can quantify important biomarkers of creatinine (CRN) and uric acid (UA) in water and synthetic urine via formation of host–guest complexes and precise plasmonic hotspots.
- The key binding parameters of the CB7-CRN and CB7-UA complexes can be quantified using NMR and UV-Vis titrations and supported by DFT molecular models.
- Aqueous self-assembly of the plasmonic substrate is dual-triggered by a combination of portal binding of CB7 and electrostatic effects of CRN.
- The nanosensor can quantify CRN and UA of clinically relevant levels in water and highly diluted SU with rapid response, high sensitivity, and reproducibility by localising CRN at the centre of the plasmonic hotspots and UA in close proximity to the hotspots.

- The nanosensor allows quantification via multiple independent spectral signatures from a single sample solution, which enhances the data consistency and measurement output.
- The CB7-Au NP SERS nanosensor is tolerant against the presence of other biomolecules and proteins in the sample matrix, making it promising for clinical applications.

References

1. Yan, S.-L., Lin, P.-Z. and Hsiao, M.-W. (1999). Separation of urea, uric acid, creatine and creatinine by micellar electrokinetic capillary chromatography with sodium cholate, *J. Chromatogr. Sci.*, **37**, pp. 45–50.
2. Premasiri, W. R., Clarke, R. H. and Womble, M. E. (2001). Urine analysis by laser Raman spectroscopy, *Lasers Surg. Med.*, **28**, pp. 330–334.
3. Chen, J.-C., Kumar, A. S., Chung, H.-H., Chien, H.-H., Kuo, M.-C. and Zen, J.-M. (2006). An enzymeless electrochemical sensor for the selective determination of creatinine in human urine, *Sens. Actuators B*, **115**, pp. 473–480.
4. Chio, W.-I. K., Moorthy, S., Perumal, J., Dinish, U. S., Parkin, I. P., Olivo, M. and Lee, T.-C. (2020). Dual-triggered nanoaggregates of cucurbit[7] uril and gold nanoparticles for multi-spectroscopic quantification of creatinine in urinalysis, *J. Mater. Chem. C*, **8**, pp. 7051–7058.
5. Chio, W.-I. K., Davison, G., Jones, T., Liu, J., Parkin, I. P. and Lee, T.-C. (2020). Quantitative SERS detection within aggregates of gold nanoparticles and cucurbit[*n*]uril, *J. Vis. Exp.*, **164**, e61682.
6. Biedermann, F., Uzunova, V. D., Scherman, O. A., Nau, W. M. and De Simone, A. (2012). Release of high-energy water as an essential driving force for high-affinity binding of cucurbit[*n*]urils, *J. Am. Chem. Soc.*, **134**, pp. 15318–15323.
7. Eisner, C., Faulhaber-Walter, R., Wang, Y., Leelahavanichkul, A., Yuen, P. S. T., Mizel, D., Star, R. A., Briggs, J. P., Levine, M. and Schnermann, J. (2010). Major contribution of tubular secretion to creatinine clearance in mice, *Kidney Int.*, **77**, pp. 519–526.
8. Li, M., Du, Y., Zhao, F., Zeng, J., Mohan, C., Shih, W.-C. (2015). Reagent- and separation-free measurements of urine creatinine concentration using stamping surface-enhanced Raman scattering (S-SERS), *Biomed. Opt. Express*, **6**, pp. 849–858.

9. Mahajan, S., Lee, T.-C., Biedermann, F., Hugall, J. T., Baumberg, J. J. and Scherman, O. A. (2010). Raman and SERS spectroscopy of cucurbit[*n*]urils, *Phys. Chem. Chem. Phys.*, **12**, pp. 10429–10433.

10. Valadbeigi, Y., Ilbeigi, V., Tabrizchi, M. (2015). Effect of mono- and di-hydration on the stability and tautomerisms of different tautomers of creatinine: a thermodynamic and mechanistic study, *Comput. Theor. Chem.*, **1061**, pp. 27–35.

11. Taylor, R. W., Lee, T.-C., Scherman, O. A., Esteban, R., Aizpurua, J., Huang, F. M., Baumberg, J. J. and Mahajan, S. (2011). Precise subnanometer plasmonic junctions for SERS within gold nanoparticle assemblies using cucurbit[*n*]uril "glue", *ACS Nano*, **5**, pp. 3878–3887.

12. Kuik, J. C., Applin, D. M., Ramirez, M. and Cloutis, E. A. (2020). Astrobiology - and habitability-relevant materials characterised by 532 nm laser Raman spectroscopy, *51st Lunar and Planetary Science Conference*, 2828.

13. Keuleers, R., Desseyn, H. O., Rousseau, B. and Van Alsenoy, C. (1999). Vibrational analysis of urea, *J. Phys. Chem. A.*, **103**, pp. 4621–4630.

14. Wang, T.-L., Chiang, H. K., Lu, H.-H and Peng, F.-Y. (2005). Semi-quantitative surface enhanced Raman scattering spectroscopic creatinine measurement in human urine samples, *Opt. Quant. Electron*, **37**, pp. 1415–1422.

15. Wang, Y., Chen, J., Wu, Y., Chen, Y., Pan, J., Lei, J., Chen, Y., Sun, L., Feng, S. and Chen, R. (2011). Surface-enhanced Raman spectroscopy of creatinine in silver colloid, *Proc. SPIE*, **8329**, 83290K.

16. Gangopadhyay, D., Sharma, P., Nandi, R., Das, M., Ghosh, S. and Singh, R. K., In vitro concentration dependent detection of creatinine: a surface enhanced Raman scattering and fluorescence study, *RSC Adv.*, **6**, 112562.

17. Wang, H., Malvadkar, N., Koytek, S., Bylander, J., Reeves, W. B. and Demirel, M. C. (2010). Quantitative analysis of creatinine in urine by metalized nanostructured parylene, *Biomed. Opt.*, **15**, 027004.

18. Zhang, H., Li, G., Li, S., Xu, L., Tian, Y., Jiao, A. and Liu, X. (2018). Boron nitride/gold nanocomposites for crystal violet and creatinine detection by surface-enhanced Raman spectroscopy, *Appl. Surf. Sci.*, **457**, pp. 684–694.

19. Lu, Y., Wu, C., You, R., Wu, Y., Shen, H., Zhu, L. and Feng, S. (2018). Superhydrophobic silver film as a SERS substrate for the detection of uric acid and creatinine, *Biomed. Opt. Express*, **9**, pp. 4988–4997.

20. Chamuah, N., Saikia, A., Joseph, A. M. and Nath, P. (2019). Blu-ray DVD as SERS substrate for reliable detection of albumin, creatinine and urea in urine, *Sen. Actuators B*, **285**, 108.
21. Westley, C., Xu, Y., Thilaganathan, B., Carnell, A. J., Turner, N. J. and Goodacre, R. (2017). Absolute quantification of uric acid in human urine using surface-enhanced Raman scattering with the standard addition method, *Anal. Chem.*, **89**, pp. 2472–2477.
22. Feig, D. I., Mazzali, M., Kang, D.-H., Nakagawa, T., Price, K., Kannelis, J. and Johnson, R. J. (2006). Serum uric acid: a risk factor and a target for treatment, *J. Am. Soc. Nephrol.*, **17**, pp. 69–73.
23. Villa, J. E. L. and Poppi, R. J. (2016). A portable SERS method for the determination of uric acid using a paper-based substrate and multivariate curve resolution, *Analyst*, **141**, pp. 1966–1972.

Chapter 7

Case Study: Methylxanthine Drugs

This chapter is intended to provide a full case study on macrocycle-functionalised nanosensors based on cucurbiturils (CB7 and CB8) and gold nanoparticles (Au NPs) for the detection of methylxanthine (MeX) drug molecules, theobromine (TBR), theophylline (TPH), and caffeine (CAF), via surface-enhanced Raman scattering (SERS), and multiplexed quantification of MeX with the aid of machine learning.

7.1 Introduction

Methylxanthines (MeX) are central nervous system stimulants that can block the adenosine receptors and inhibit phosphodiesterases to sustain alertness with antitumour and anti-inflammatory properties [1–3]. Theobromine (TBR, 3,7-dimethylxanthine) and theophylline (TPH, 1,3-dimethylxanthine) are active ingredients of bronchodilator drugs for respiratory diseases such as asthma while caffeine (CAF, 1,3,7-trimethylxanthine) is commonly used in pharmaceutical formulation. The structural isomers co-exist as major metabolites of CAF in blood plasma and urine [4]. MeX are also naturally present in foods and beverages including chocolate, tea, and coffee, which are potentially toxic to small animals and

Macrocycle-Functionalised Nanosensors
Weng-I Katherine Chio and Tung-Chun Lee

ISBN 978-981-5129-33-5 (Hardcover), 978-1-003-51057-4 (eBook)
www.jennystanford.com

even humans in the case of severe overdose. Conventional sensing methods for MeX in therapeutic drug monitoring and toxicity testing are based on high-performance liquid chromatography (HPLC) and enzyme immunoassay techniques. The former is time-consuming and requires expensive instruments while the latter is moderately sensitive, expensive, and prone to interferences with structurally similar compounds. Hence, it is important to develop a high-performance MeX sensor with multiplexing ability for disease diagnosis, therapeutic drug monitoring, and quality control of consumer products (Fig. 7.1).

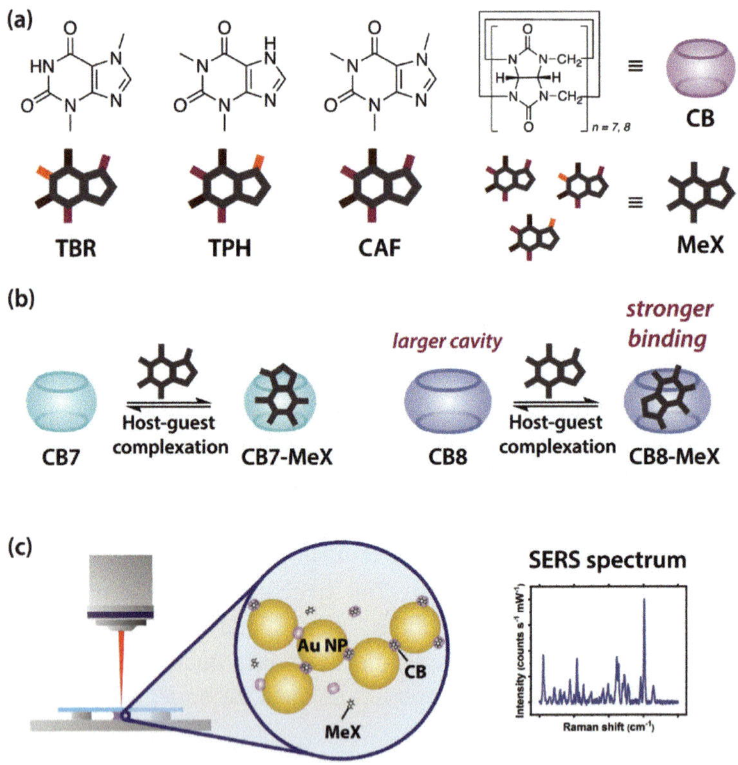

Figure 7.1 Schematic illustrations of (a) MeX (including TBR, TPH, and CAF) and CB, (b) the formation of 1:1 CB7-MeX and CB8-MeX host–guest complexes, and (c) the SERS measurement of the CB-Au NP-MeX solution.

7.2 CB7-Au NP and CB8-Au NP Nanosensors for Methylxanthine Drugs

Macrocycle-functionalised nanosensors can detect and distinguish structurally similar analyte molecules, which are encapsulated within the cavity of the macrocycle and thus localised at the centre of or in close proximity to the plasmonic hotspots, via surface-enhanced Raman scattering (SERS). The SERS nanosensor can be coupled to advanced data analysis techniques for multiplexed detection of analyte molecules with additional advantages such as high sensitivity, rapid response, cost-effectiveness, and reproducibility. This section will discuss a case on sensing of structurally similar MeX drug molecules with nanoaggregates of cucurbit[*n*]urils (CB*n*, *n* = 7 or 8) and gold nanoparticles (Au NPs), covering from formation of host–guest complexes and plasmonic nanojunctions, through quantification of different MeX molecules, to multiplexed detection of isomers using machine learning [5].

7.2.1 Supramolecular Chemistry of CB and MeX

The conjugated acids of TBR, TPH, and CAF, i.e., $[TBR\text{-}H]^+$, $[TPH\text{-}H]^+$, and $[CAF\text{-}H]^+$, have pKa of 10.0, 8.8, and 10.44, respectively, implying that they exist in the protonated form in water [2]. The formation of 1:1 host–guest complexes between CB7 and TBR can be verified by the characteristic upfield shifts of the TBR proton signals (H_b and H_c) in the ^1H NMR spectra upon mixing in D_2O (Fig. 7.2b). Note that the change in chemical shift ($\Delta\delta$) for H_b is significant (−0.099 ppm) and the NMR signal is broadened. This implies H_b is located deep inside the CB7 cavity, and the binding kinetics fall into the intermediate exchange regime on the NMR timescale at 298 K. It can be challenging to observe the very broad H_a signal in NMR spectra. In addition, the larger CB homologue, CB8, can also form 1:1 host–guest complexes with TBR in solution, as verified by the characteristic upfield shifts of the TBR proton signals (H_a, H_b, and H_c) in the NMR spectra after mixing CB8 and TBR in 10 mM DCl (Fig. 7.3b). In particular, $\Delta\delta$ for H_a (−0.014 ppm) is greater than that for H_c/H_b (−0.010 ppm/−0.011 ppm), implying that H_a is deeper inside the CB8 cavity while H_c/H_b

are closer to the carbonyl portals. The relatively smaller upfield shift is due to the weaker shielding effect of the CB8 cavity. Note that all proton signals are in the fast exchange regime and thus consistent with the larger portal and less constrictive binding of CB8.

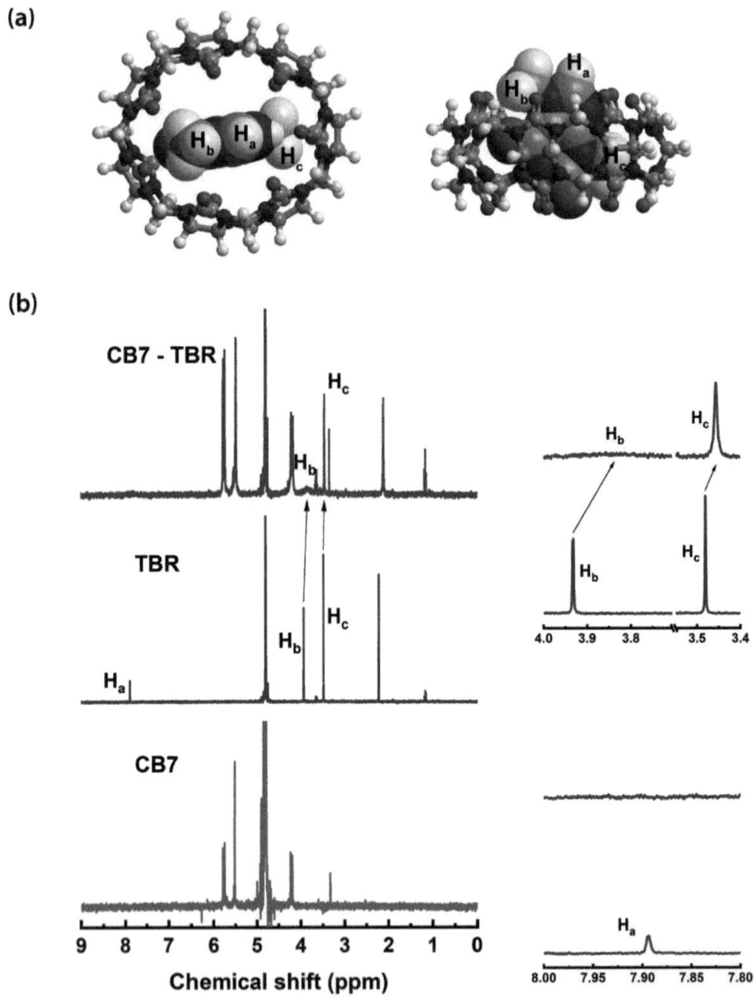

Figure 7.2 (a) DFT molecular model of a $[CB7\text{-}TBR\text{-}H]^+$ host–guest complex in water. (b) Full-range and zoom-in ^1H NMR spectra of CB7, TBR, and 1:1 CB7-TBR host–guest complexes in D_2O.

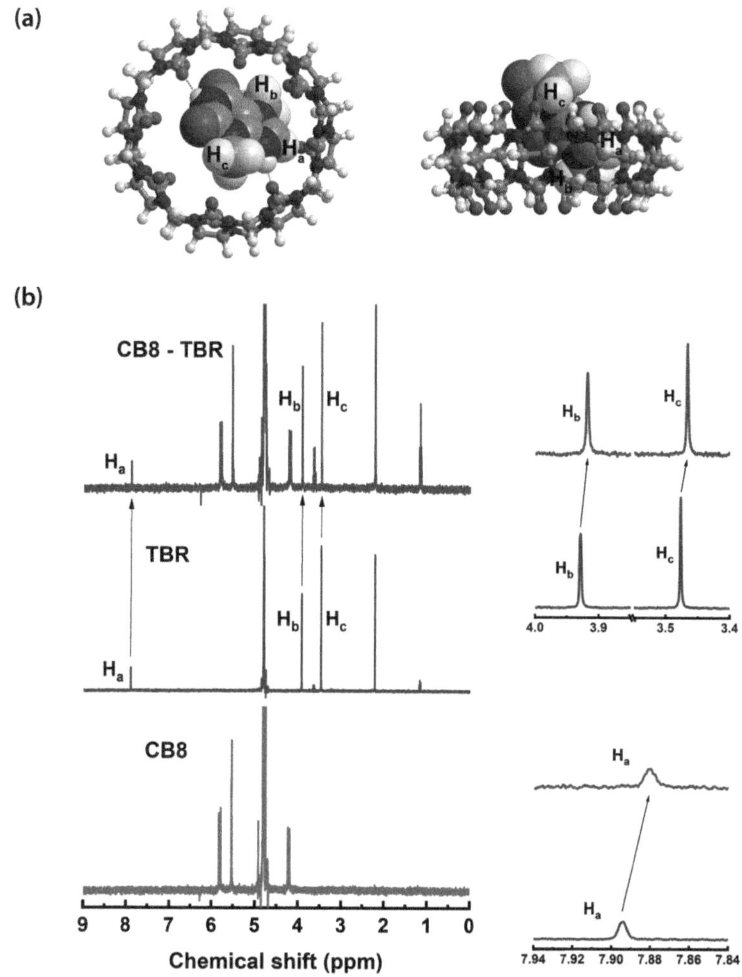

Figure 7.3 (a) DFT molecular model of a [CB8-TBR-H]$^+$ host–guest complex in water. (b) Full-range and zoom-in ^{1}H NMR spectra of CB8, TBR, and 1:1 CB8-TBR host–guest complexes in DCl.

On the other hand, the formation of host–guest complexes between CB and TPH can also be verified by the characteristic shifts of the TPH proton signals (H_a, H_b, and H_c) in ^{1}H NMR upon mixing CB and TPH with 1:1 stoichiometry (Figs. 7.4b and 7.5b). In both cases of CB7-TPH and CB8-TPH, H_c/H_b are upfield shifted while H_a is downfield shifted after host–guest complexation, indicating that H_c/H_b are within the CB7 cavity while H_a is around the portal region. Note that there is no strong broadening of proton signals in the case of TPH even though TBR and TPH are structural isomers.

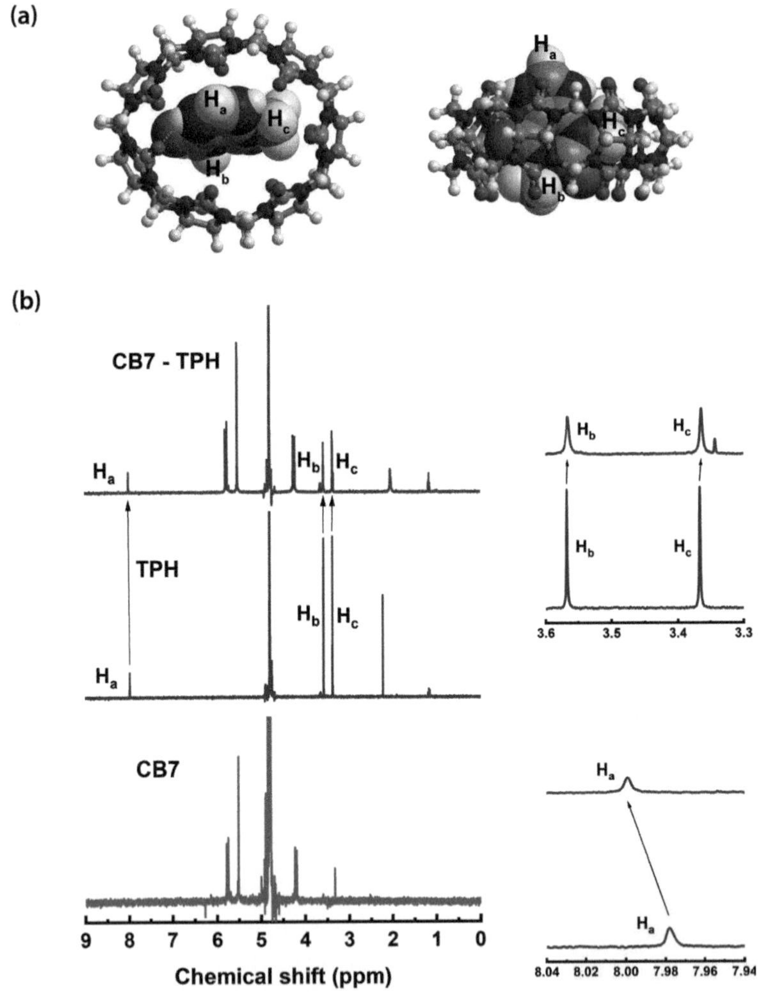

Figure 7.4 (a) DFT molecular model of a $[CB7\text{-}TPH\text{-}H]^+$ host–guest complex in water. (b) Full-range and zoom-in ^1H NMR spectra of CB7, TPH, and 1:1 CB7-TPH host–guest complexes in D_2O.

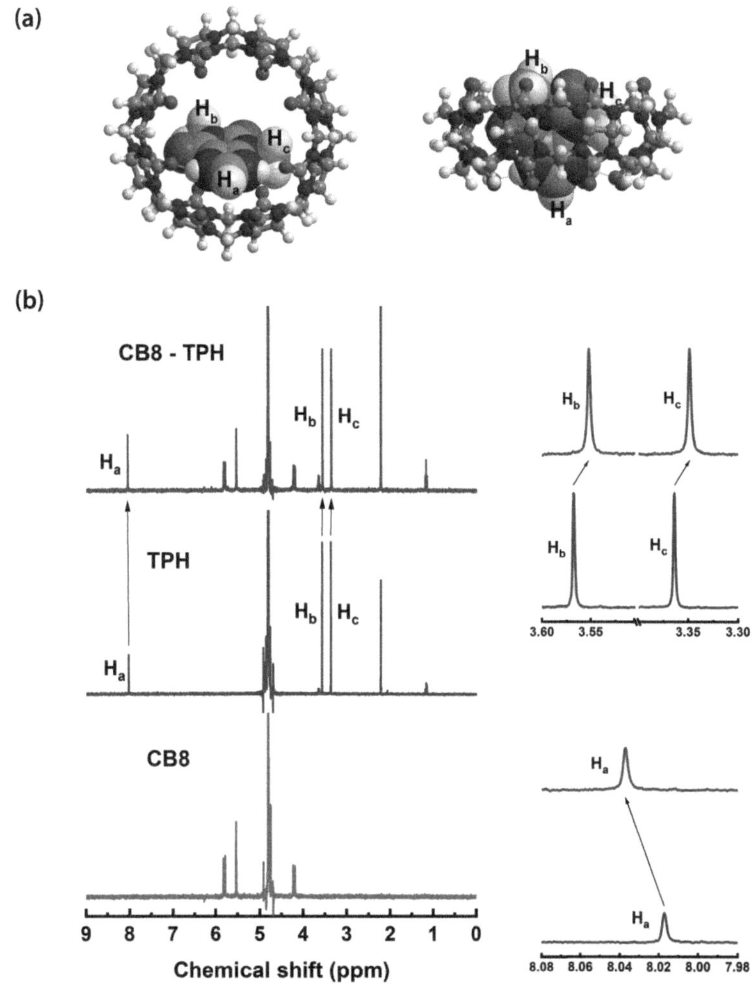

Figure 7.5 (a) DFT molecular model of a [CB8-TPH-H]$^+$ host–guest complex in water. (b) Full-range and zoom-in ^{1}H NMR spectra of CB8, TPH, and 1:1 CB8-TPH host–guest complexes in DCl.

Compared to TBR and TPH, the relatively bulky CAF binds weakly to CB7, as evidenced by the smaller upfield shifts of the proton signals in NMR (Fig. 7.6b). Note that the NMR signal of H_d splits upon the formation of 1:1 CB7-CAF host–guest complexes, resulting in two distinct values of $\Delta\delta$ (–0.002 ppm and –0.010 ppm). $\Delta\delta$ for H_d

is greater than that for $H_c/H_b/H_a$ (−0.002 ppm/−0.002 ppm/−0.003 ppm), implying H_d is deep inside the cavity of CB7, whereas the others are much closer to the carbonyl portal. In addition, the CB8-CAF complexation is similar to that of CB7-CAF despite the absence of signal splitting (Fig. 7.7b).

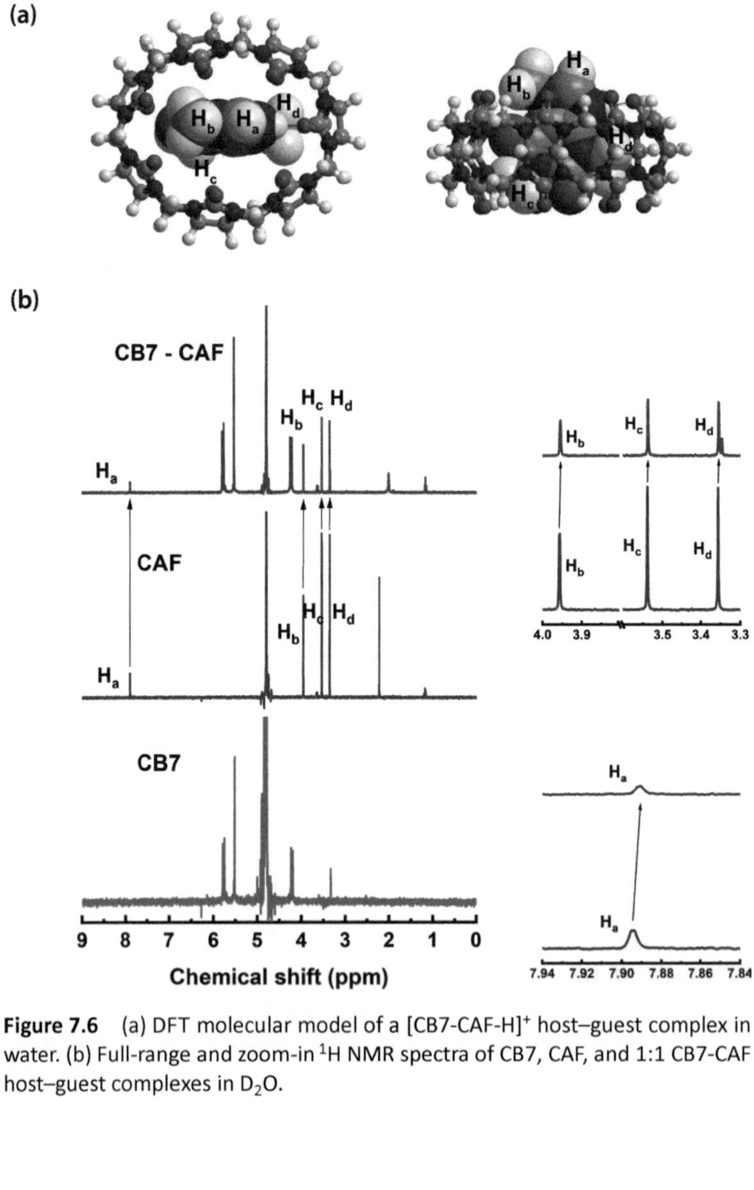

Figure 7.6 (a) DFT molecular model of a $[CB7\text{-}CAF\text{-}H]^+$ host–guest complex in water. (b) Full-range and zoom-in ^1H NMR spectra of CB7, CAF, and 1:1 CB7-CAF host–guest complexes in D_2O.

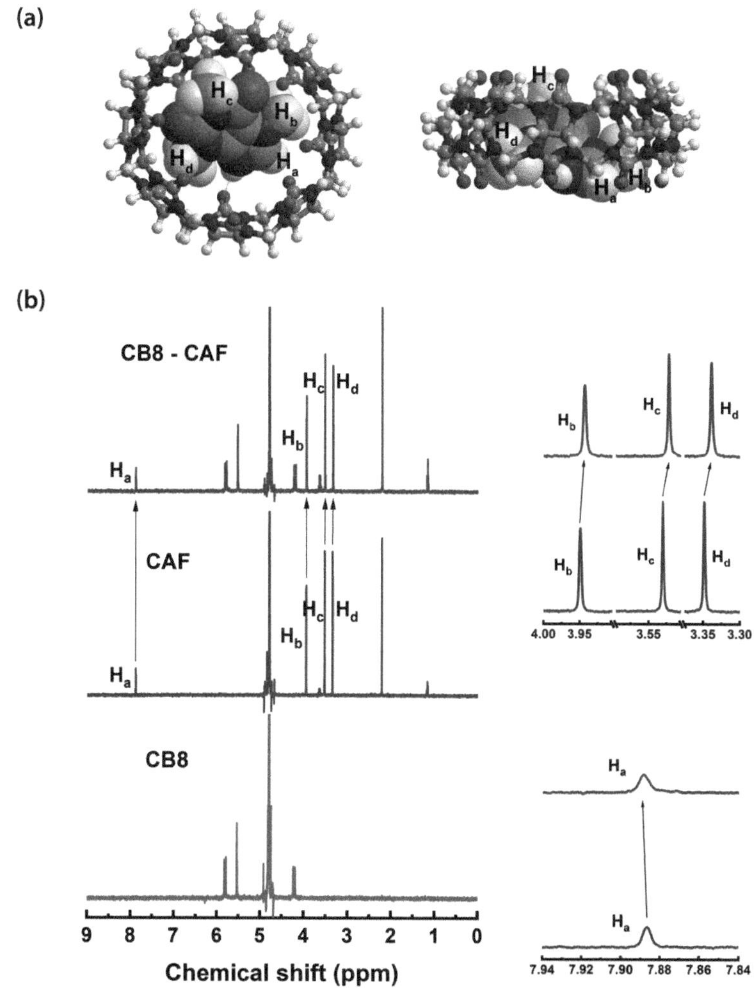

Figure 7.7 (a) DFT molecular model of a [CB8-CAF-H]$^+$ host–guest complex in water. (b) Full-range and zoom-in ^{1}H NMR spectra of CB8, CAF, and 1:1 CB8-CAF host–guest complexes in DCl.

Binding constants of the CB-MeX complexes can be obtained from UV-Vis titrations by fitting a 1:1 binding model (Fig. 7.8). Note that the binding constants of CB8-TBR (1.05×10^5 M^{-1}) and CB8-TPH (7.35×10^5 M^{-1}) are fivefold and 20-fold larger than that for CB7-TBR (2.08×10^4 M^{-1}) and CB7-TPH (3.85×10^4 M^{-1}), respectively, while the binding constants of CB7-CAF (5.83×10^4 M^{-1}) and CB8-

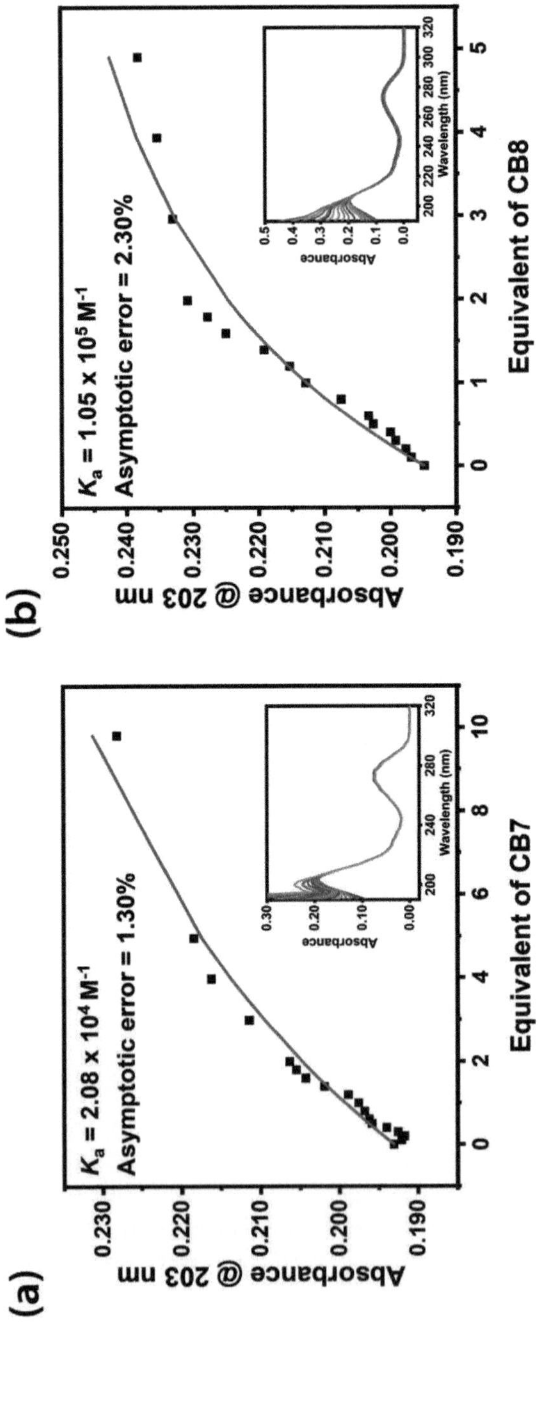
(a)
K_a = 2.08 x 10^4 M^{-1}
Asymptotic error = 1.30%
Absorbance @ 203 nm
Equivalent of CB7
Absorbance
Wavelength (nm)
(b)
K_a = 1.05 x 10^5 M^{-1}
Asymptotic error = 2.30%
Absorbance @ 203 nm
Equivalent of CB8
Absorbance
Wavelength (nm)

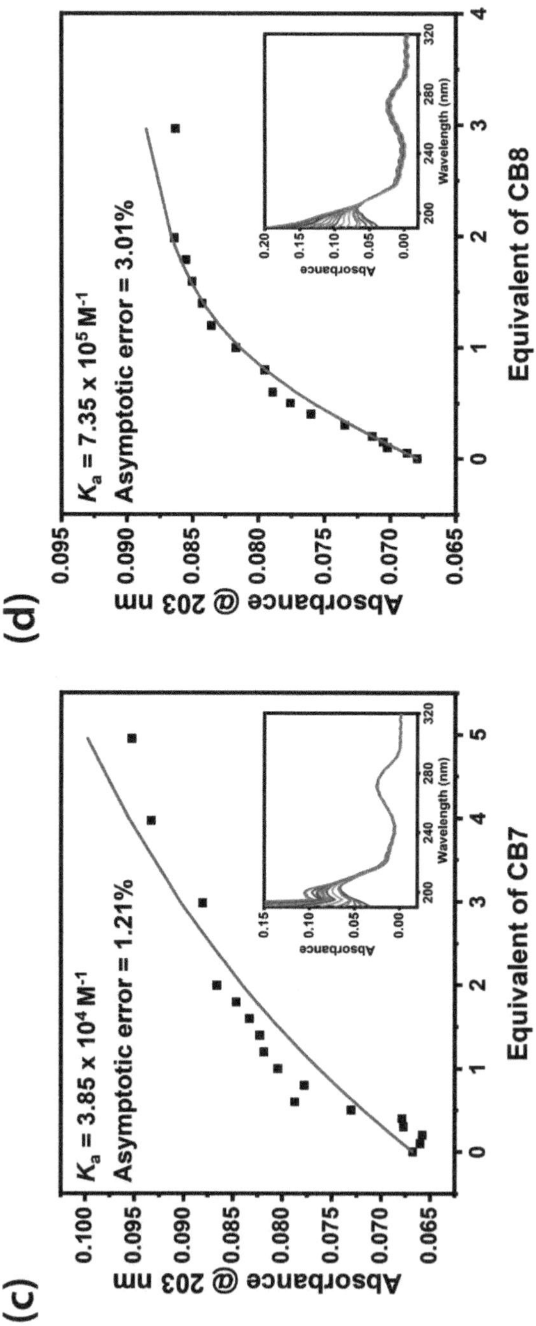
(c)
K_a = 3.85 x 10^4 M^{-1}
Asymptotic error = 1.21%
Absorbance @ 203 nm
Equivalent of CB7
Absorbance
Wavelength (nm)
(d)
K_a = 7.35 x 10^5 M^{-1}
Asymptotic error = 3.01%
Absorbance @ 203 nm
Equivalent of CB8
Absorbance
Wavelength (nm)

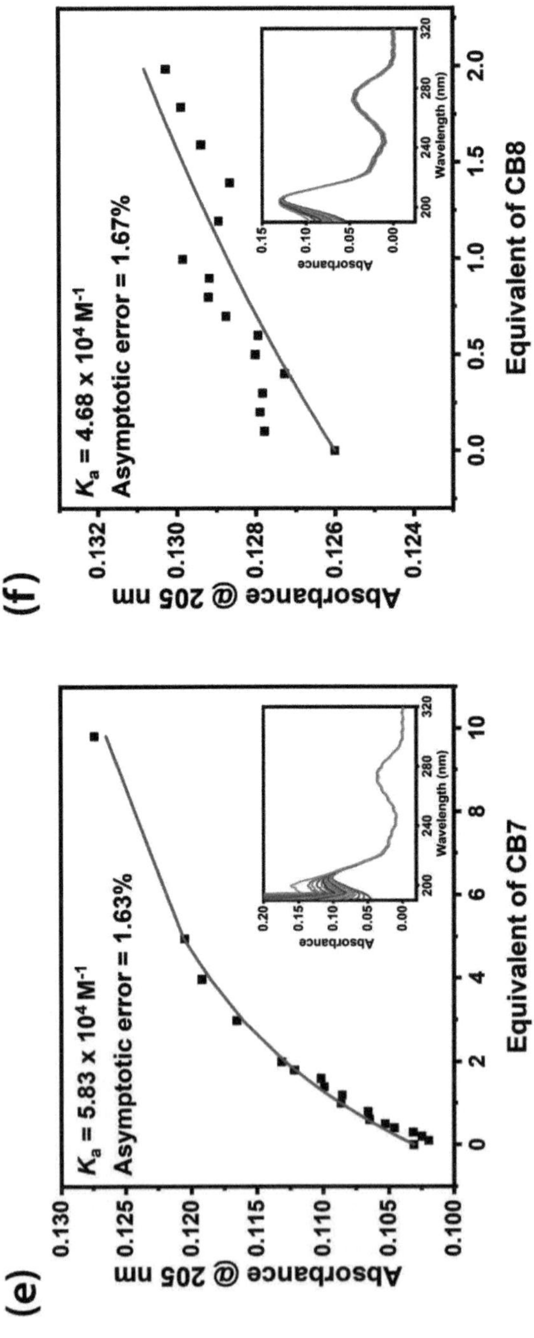

Figure 7.8 UV-Vis titration of 4 μM (a) TBR with CB7, (b) TBR with CB8, (c) TPH with CB7, (d) TPH with CB8, (e) CAF with CB7, and (f) CAF with CB8. Insets: UV-Vis spectra of TBR, TPH, or CAF upon stepwise addition of CB7 or CB8.

CAF (4.68×10^4 M^{-1}) are similar. The larger binding constants of CB8-TBR and CB8-TPH compared to their CB7 counterparts can be explained by the host–guest size matching phenomenon, the Rebek's 55% rule, which states a packing coefficient (PC) of 55% gives the optimal binding affinity for host–guest complexes in solution and that a lower or higher PC can lead to a lower binding affinity [6]. The packing coefficients of CB8-TBR and CB8-TPH are 46% while those of CB7-TBR and CB7-TPH are 70% [5]. Also, it can be visualised from the molecular models that CB8 retains a circular confirmation upon host–guest complexation whereas CB7 is highly distorted and conformationally destabilised (Figs. 7.2a–7.7a). Note that it can be challenging to minimise the fitting error owing to the interference from the UV absorption of CBs and turbidity of CB8 at high equivalents.

Energy-minimised molecular models of the $[CB7\text{-}MeX\text{-}H]^+$ and $[CB8\text{-}MeX\text{-}H]^+$ complexes and their binding energies in water (Table 7.1) can be calculated by DFT at CPCM/wB97X-D/6-31G* level of theory. In particular, CPCM is an implicit water model used to account for the solvation energy of the protonated complexes with a formal charge of +1, while wB97X-D is a dispersion-corrected DFT functional selected to estimate the van der Waals interactions contributed significantly to the stability of the complexes. The DFT molecular models support the host–guest binding geometries in solution deduced from NMR (Fig. 7.2a–7.7a). MeX mostly reside in the CB cavity, typically with the methyl group occupying the portal region, which could lead to a slightly weakened interaction between CB and Au NPs in the nanoaggregates. Note that MeX do not protrude the cavity entirely, like viologen or 2,4-dinitrotoluene (see Section 5.2.1), leaving at least one of the CB portals available for binding to the Au NPs and thus localising the analyte molecules at the centre of or in close proximity to the plasmonic hotspots for SERS sensing.

Table 7.1 Binding energies of the $[CB\text{-}MeX\text{-}H]^+$ complexes in water at the CPCM/wB97X-D/6-31G* level of theory

Binding energy/kcal mol^{-1}			
$[CB7\text{-}TBR\text{-}H]^+$	-39.0066	$[CB8\text{-}TBR\text{-}H]^+$	-34.5581
$[CB7\text{-}TPH\text{-}H]^+$	-43.3128	$[CB8\text{-}TPH\text{-}H]^+$	-41.4718
$[CB7\text{-}CAF\text{-}H]^+$	-40.3841	$[CB8\text{-}CAF\text{-}H]^+$	-39.1747

7.2.2 Raman Spectroscopy of MeX and SERS of CB

TBR powder can be characterised by two major peaks at 622 cm^{-1} and 1334 cm^{-1} in the Raman spectrum, corresponding to C═C—C deformation and imidazole ring stretching vibration, respectively (Fig. 7.9a) [7, 8]. TPH power can be characterised by a main peak at 557 cm^{-1}, which is assigned to pyrimidine ring deformation + C—N—C deformation + CH_3 rocking (Fig. 7.9b), together with two other peaks attributed to O═C—N deformation + pyrimidine imidazole ring deformation and imidazole ring stretching vibration at 668 cm^{-1} and 1316 cm^{-1}, respectively [7]. On the other hand, the larger member of MeX, CAF powder can be characterised by two major peaks at 558 cm^{-1} and 1330 cm^{-1}, which can be attributed to pyrimidine ring deformation + C—N—C deformation + CH_3 rocking and imidazole ring stretching vibration, respectively (Fig. 7.9c) [7, 9]. The similar molecular structures of MeX give rise to the analogous Raman spectra.

While CB7 crystals can be characterised by two main peaks at 444 cm^{-1} and 833 cm^{-1} (see Section 5.2.2), the former shifts slightly in position to 447 cm^{-1}, and both peaks broaden upon addition to Au NP solution. This is due to a combination of molecular interactions between CB7 and Au NPs, and solution effects (Fig. 7.9d). Note that CB7 and CB8 can be distinguished by the position of the two characteristic CB peaks, with the ring scissor mode of the larger homologue at a lower wavenumber of 444 cm^{-1} and the ring deformation mode at a slightly higher wavenumber of 834 cm^{-1} [10].

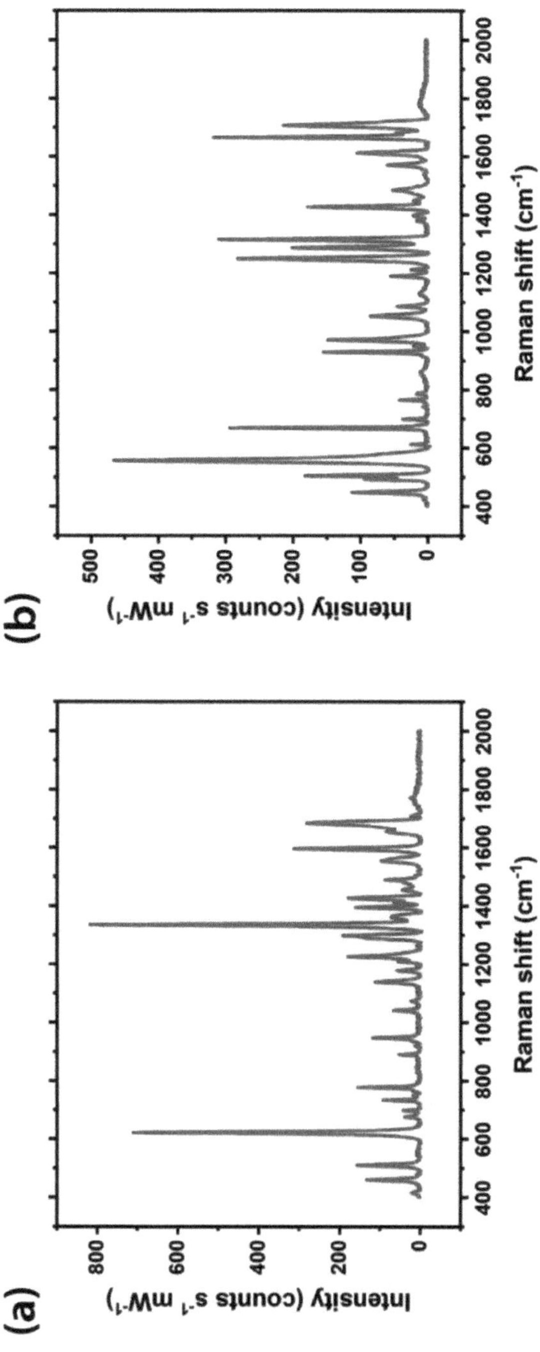
(a)
Intensity (counts s-1 mW-1)
800
600
400
200
0
400
600
800
1000
1200
1400
1600
1800
2000
Raman shift (cm-1)
(b)
Intensity (counts s-1 mW-1)
500
400
300
200
100
0
400
600
800
1000
1200
1400
1600
1800
2000
Raman shift (cm-1)

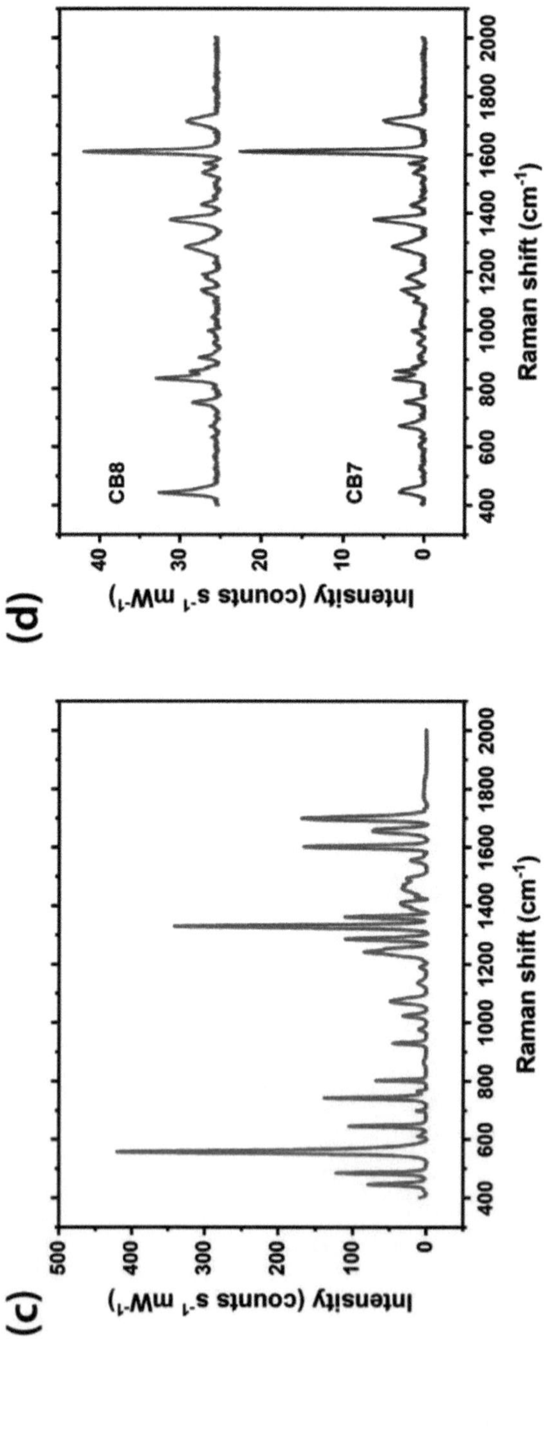

Figure 7.9 (a–c) Raman spectra of TBR, TPH, and CAF powder, respectively. (d) SERS spectra of 10 μM CB7 and 5 μM CB8.

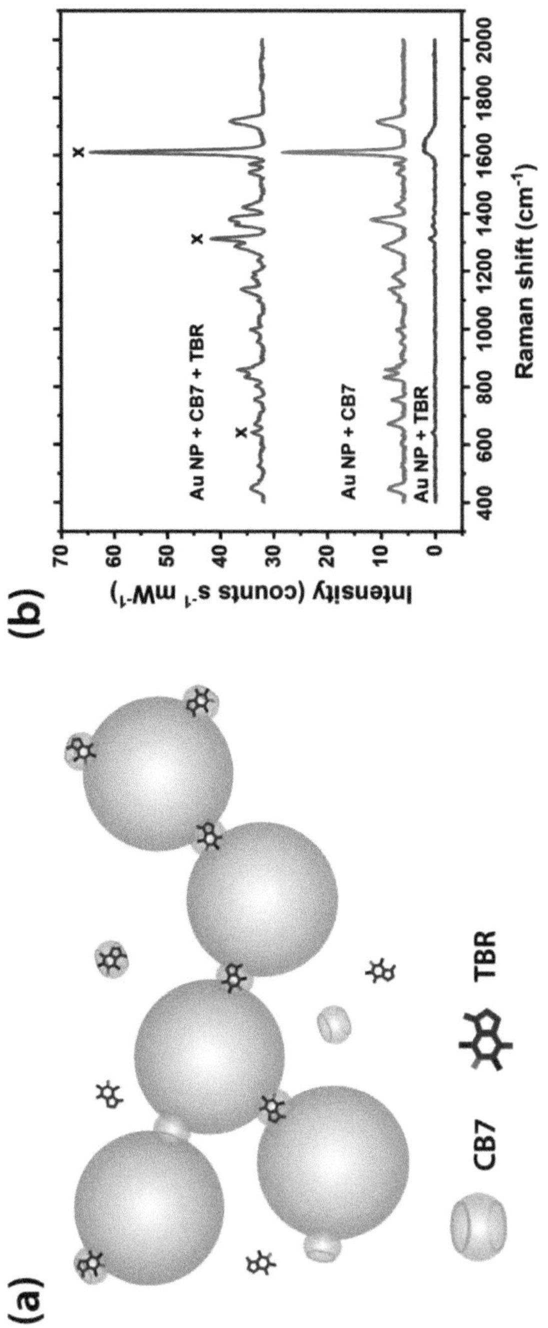
(a)
CB7
TBR
(b)
Intensity (counts s^{-1} mW^{-1})
Raman shift (cm^{-1})
Au NP + CB7 + TBR
Au NP + CB7
Au NP + TBR
x
x
x

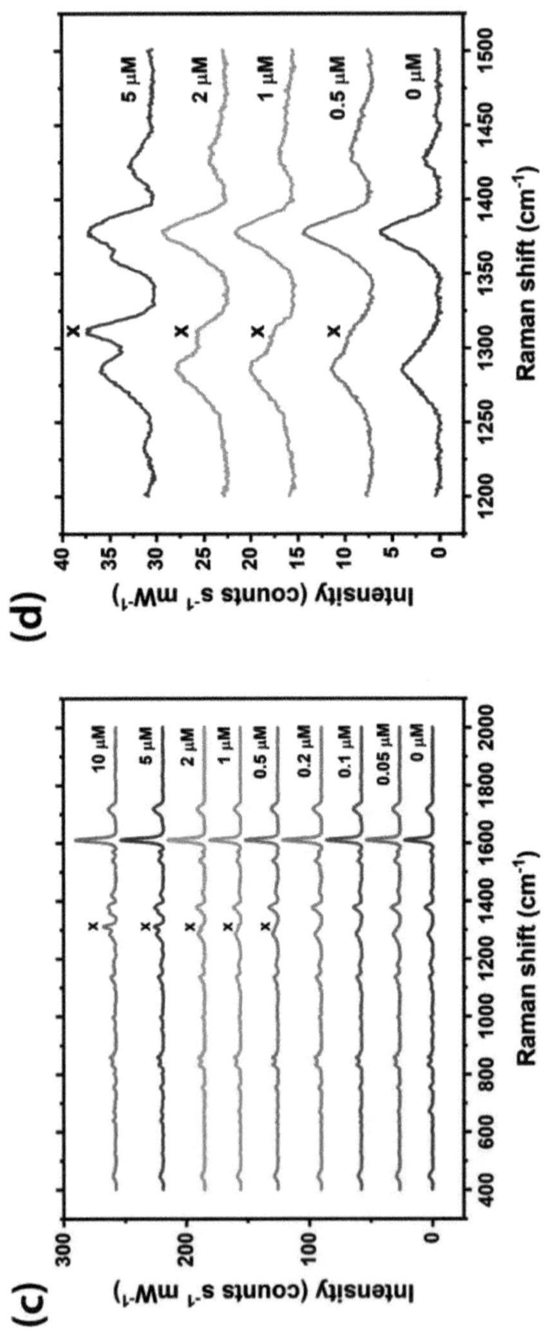
(c)
Intensity (counts s⁻¹ mW⁻¹)
Raman shift (cm⁻¹)
10 μM
5 μM
2 μM
1 μM
0.5 μM
0.2 μM
0.1 μM
0.05 μM
0 μM
(d)
Intensity (counts s⁻¹ mW⁻¹)
Raman shift (cm⁻¹)
5 μM
2 μM
1 μM
0.5 μM
0 μM

(e)

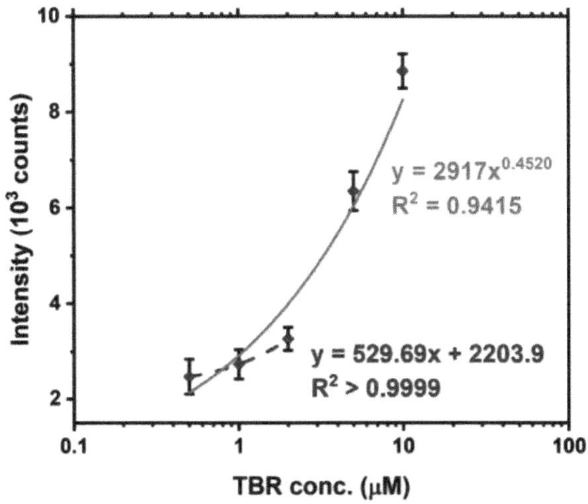

Figure 7.10 (a) Schematic illustration of the precise plasmonic hotspots within the CB7-Au NP nanosensor for TBR detection (not to scale). (b) SERS spectra of TBR with or without CB7. (c) Full-range and (d) zoom-in SERS spectra of TBR with concentrations from 0 to 10 μM. Main Raman peak of TBR at 1312 cm^{-1} is marked by x. (e) Corresponding plot of SERS intensity of the main TBR peak in (d) against its concentration.

7.2.3 Sensitivity of the SERS Nanosensor for MeX

MeX can be detected using the CB-Au NP nanosensor by adding a pre-mixed CB-MeX solution into the Au NP solution (Figs. 7.10a–7.15a). A concentration of 10 μM CB7 or 5 μM CB8 is used to generate precise plasmonic hotspots and ensure the reproducibility of the SERS sensors. As MeX mostly reside in the CB cavity (see Section 7.2.1), they are localised at the centre of or in close proximity to the plasmonic hotspots within the CB-Au NP nanoaggregates for SERS sensing. Note that no aggregation can be triggered without CB, hence illustrating the importance of the macrocycle on MeX detection (Figs. 7.10b–7.15b).

The detection limit of the CB7-Au NP nanosensor for TBR can be tested by performing SERS titrations, in which the characteristic Raman peak of TBR at 1312 cm^{-1} can be clearly observed down to 0.5 μM (Fig. 7.10c,d). A good correlation ($R^2 \sim 0.94$) between the

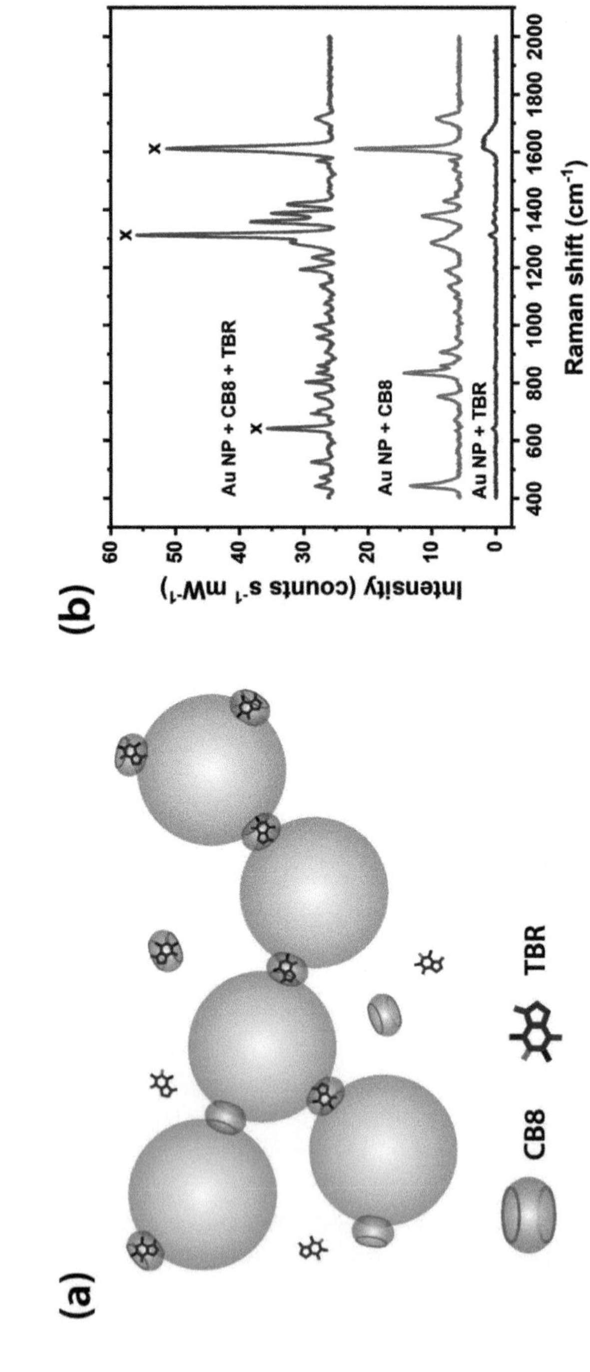
(a)
CB8
TBR
(b)
Au NP + CB8 + TBR
Au NP + CB8
Au NP + TBR
Intensity (counts s-1 mW-1)
Raman shift (cm-1)
0
10
20
30
40
50
60
400
600
800
1000
1200
1400
1600
1800
2000

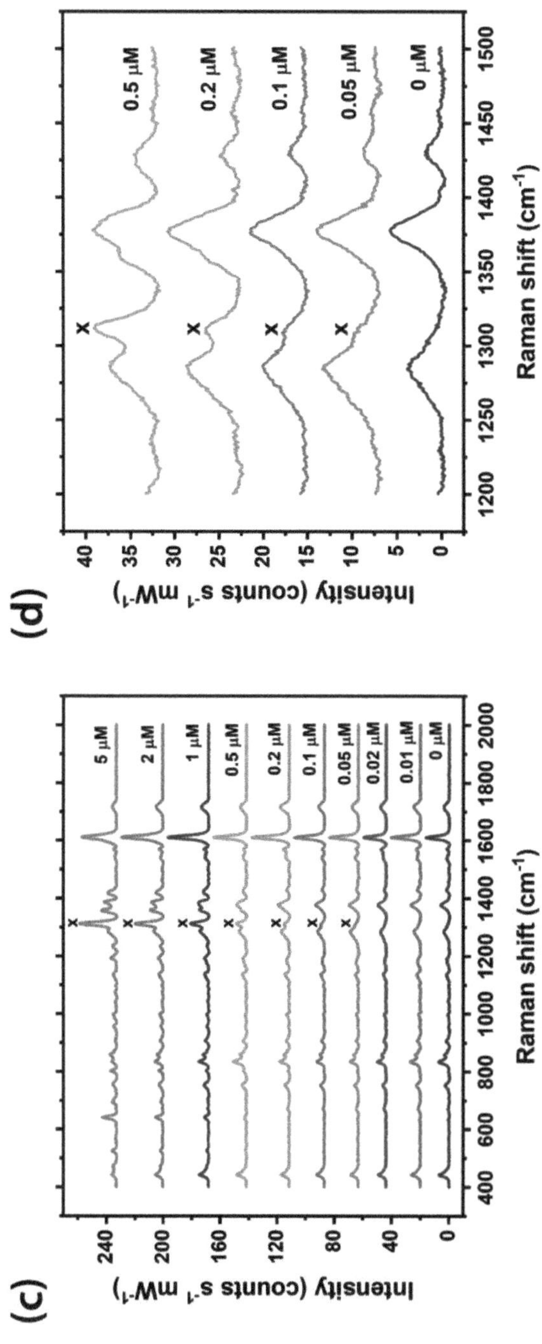
(c)
Intensity (counts s-1 mW-1)
Raman shift (cm-1)
5 μM
2 μM
1 μM
0.5 μM
0.2 μM
0.1 μM
0.05 μM
0.02 μM
0.01 μM
0 μM
(d)
Intensity (counts s-1 mW-1)
Raman shift (cm-1)
0.5 μM
0.2 μM
0.1 μM
0.05 μM
0 μM

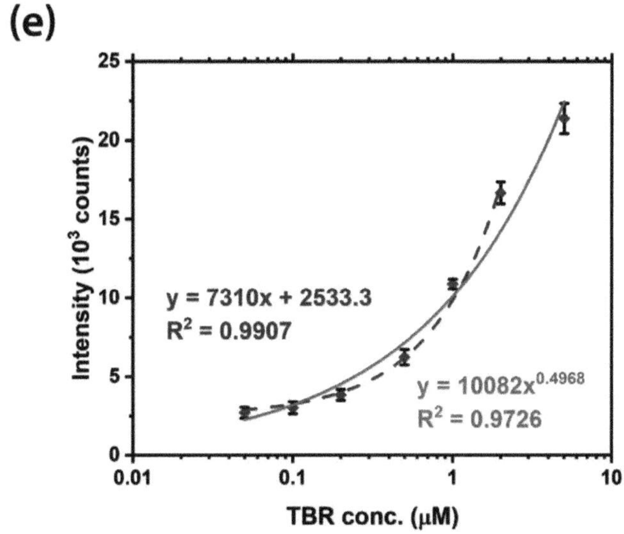

Figure 7.11 (a) Schematic illustration of the precise plasmonic hotspots within the CB8-Au NP nanosensor for TBR detection (not to scale). (b) SERS spectra of TBR with or without CB8. (c) Full-range and (d) zoom-in SERS spectra of TBR with concentrations from 0 to 5 μM. Main Raman peak of TBR at 1312 cm^{-1} is marked by x. (e) Corresponding plot of SERS intensity of the main TBR peak in (d) against its concentration.

SERS intensity and concentration of TBR can be obtained by power law while a strong linear correlation ($R^2 > 0.99$) can be fitted at low concentration of TBR up to 2 μM (Fig. 7.10e). The sensitivity of macrocycle-functionalised nanosensors for TBR detection can be significantly enhanced by using the larger CB8. Note that 10 mM HCl can be added to the pre-mixed CB8-TBR solution for better dissolution of CB8. The detection limit of the CB8-Au NP nanosensor for TBR is down to ~50 nM, which is 10-fold lower than that of the CB7-Au NP nanosensor (Fig. 7.11c,d). There is a strong linear correlation ($R^2 \sim 0.99$) between the SERS intensity and the concentration of TBR up to 2 μM with a good correlation ($R^2 \sim 0.97$) for the full range via power-law fitting (Fig. 7.11e).

On the other side, TPH, the structural isomer of TBR, can be detected down to 50 nM using the CB7-Au NP nanosensor, in which the characteristic Raman peak of TPH at 573 cm^{-1} can be clearly observed (Fig. 7.12c,d) with a very strong linear correlation ($R^2 \sim 0.99$) between the SERS intensity and the concentration of

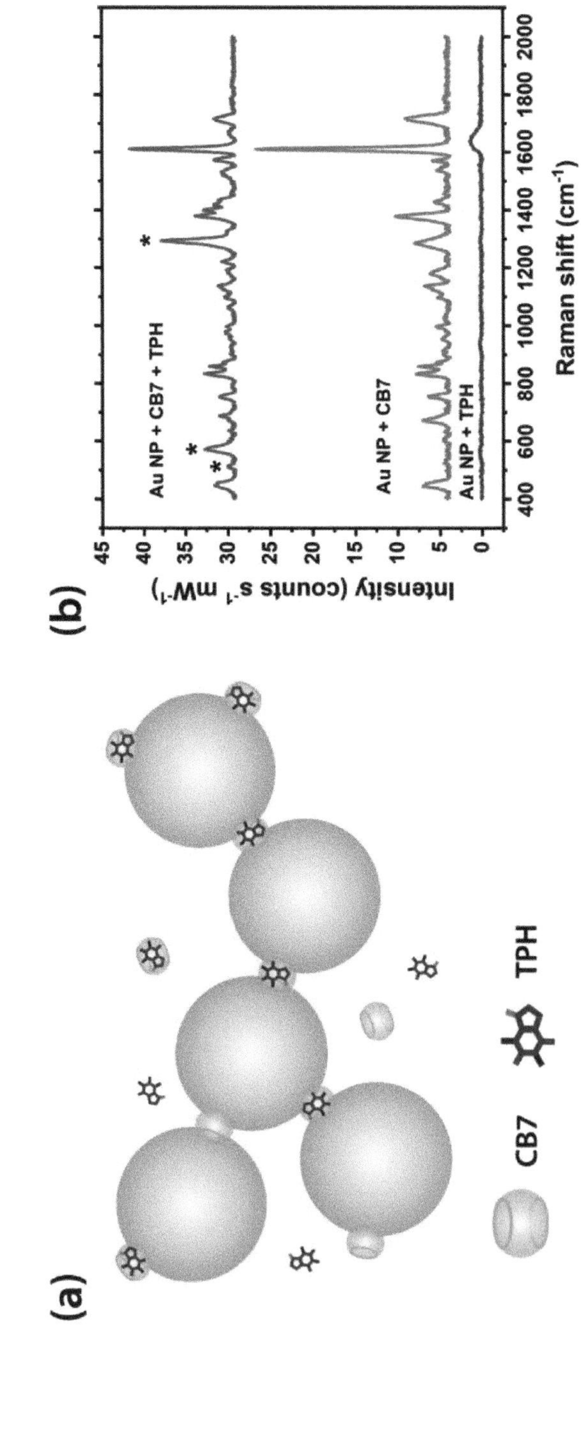
(a)
CB7
TPH
(b)
Intensity (counts s^{-1} mW^{-1})
Raman shift (cm^{-1})
Au NP + CB7 + TPH
Au NP + CB7
Au NP + TPH
0 5 10 15 20 25 30 35 40 45
400 600 800 1000 1200 1400 1600 1800 2000

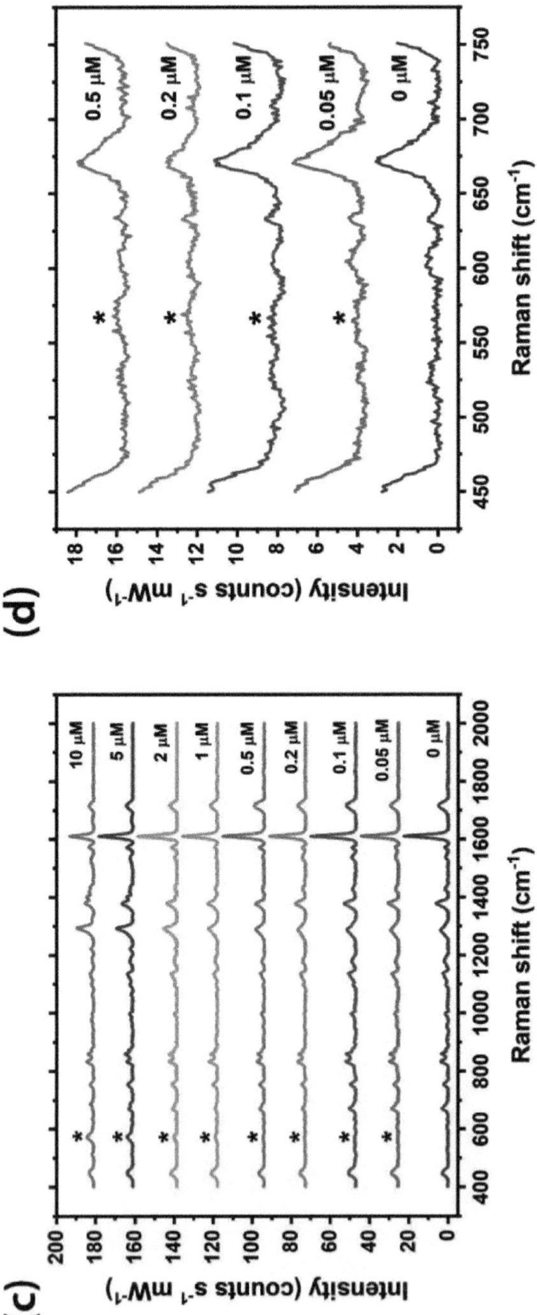
(c)
Intensity (counts s^-1 mW^-1)
Raman shift (cm^-1)
10 μM
5 μM
2 μM
1 μM
0.5 μM
0.2 μM
0.1 μM
0.05 μM
0 μM
(d)
Intensity (counts s^-1 mW^-1)
Raman shift (cm^-1)
0.5 μM
0.2 μM
0.1 μM
0.05 μM
0 μM

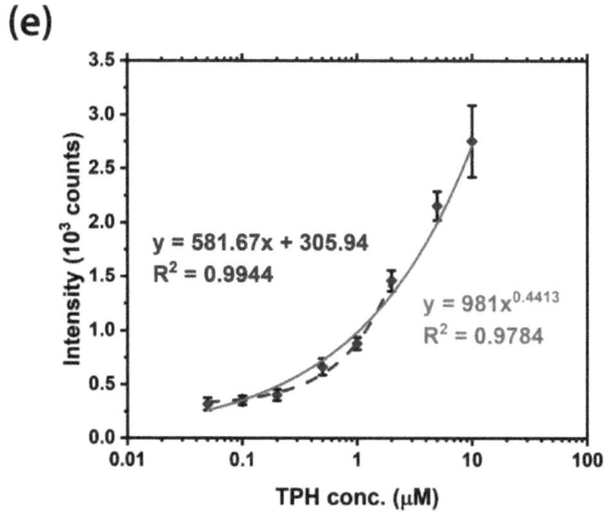

Figure 7.12 (a) Schematic illustration of the precise plasmonic hotspots within the CB7-Au NP nanosensor for TPH detection (not to scale). (b) SERS spectra of TPH with or without CB7. (c) Full-range and (d) zoom-in SERS spectra of TPH with concentrations from 0 to 10 μM. Main Raman peak of TPH at 557 cm^{-1} is marked by *. (e) Corresponding plot of SERS intensity of the main TPH peak in (d) against its concentration.

TPH up to 2 μM and a good correlation (R^2 ~ 0.98) for the full range fitted by power law (Fig. 7.12e). Despite being less Raman-active, the detection limit of TPH is 10-fold better than that of TBR using the CB7-Au NP nanosensor owing to the different binding geometries. In addition, the sensitivity of the CB8-Au NP nanosensor for TPH is similar to that of the CB7-Au NP nanosensor with a detection limit of ~0.1 μM and a strong linear correlation (R^2 ~ 0.98) between the SERS intensity and the concentration of TPH from 0.1 to 2 μM (Fig. 7.13c–e), which matches with the similar binding geometries of the $[CB8\text{-}TPH\text{-}H]^+$ and $[CB7\text{-}TPH\text{-}H]^+$ complexes. Note that the data point of 5 μM TPH deviates from the trendline, it could be due to the hindered aggregation at higher equivalence of TPH, potentially shifting the LSPR peak away from 633 nm laser excitation. On account of the strong surface enrichment effect of CBs, the detection limits of TBR and TPH using the CB-Au NP nanosensors are at least an order of magnitude better than similar SERS sensors (Table 7.2) meanwhile covering the lower bound of MeX concentration in human urine samples [2].

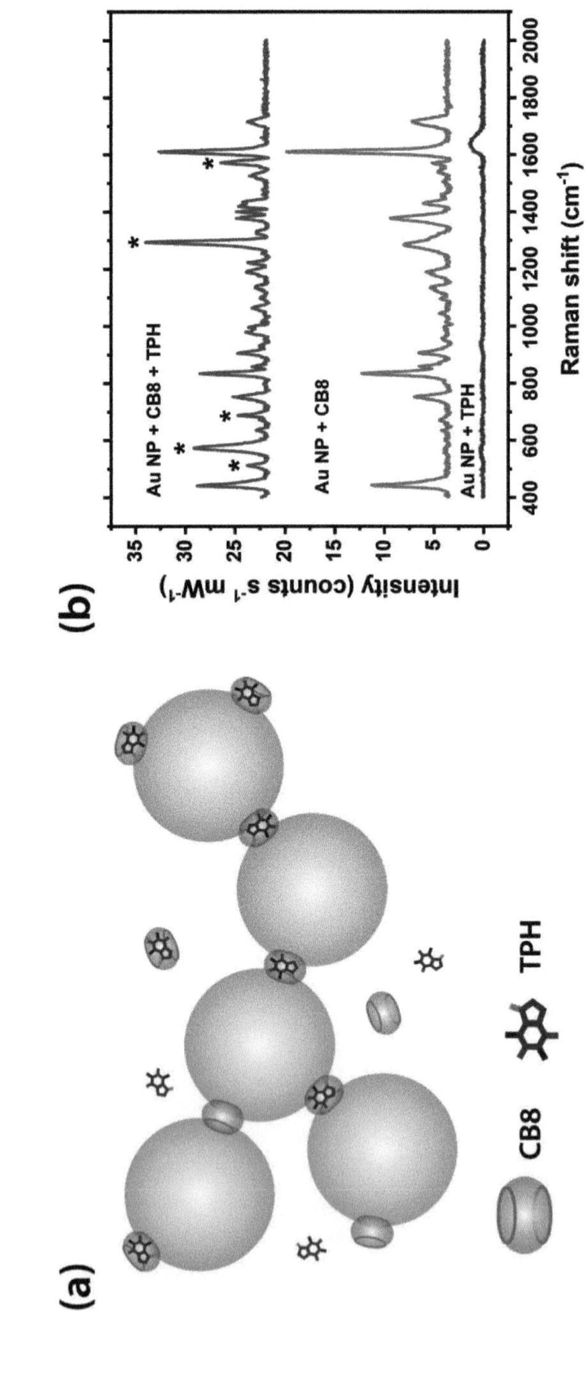
(a)
CB8
TPH
(b)
Au NP + CB8 + TPH
Au NP + CB8
Au NP + TPH
Raman shift (cm-1)
Intensity (counts s-1 mW-1)

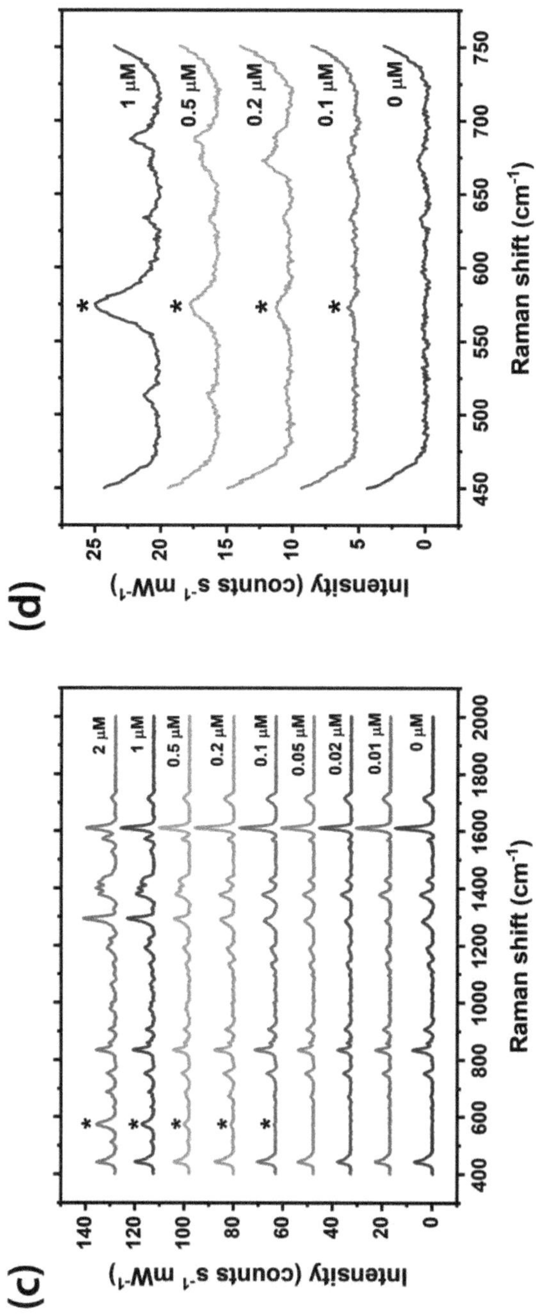
(c)
Intensity (counts s^{-1} mW^{-1})
Raman shift (cm^{-1})
2 μM
1 μM
0.5 μM
0.2 μM
0.1 μM
0.05 μM
0.02 μM
0.01 μM
0 μM
(d)
Intensity (counts s^{-1} mW^{-1})
Raman shift (cm^{-1})
1 μM
0.5 μM
0.2 μM
0.1 μM
0 μM

(e)

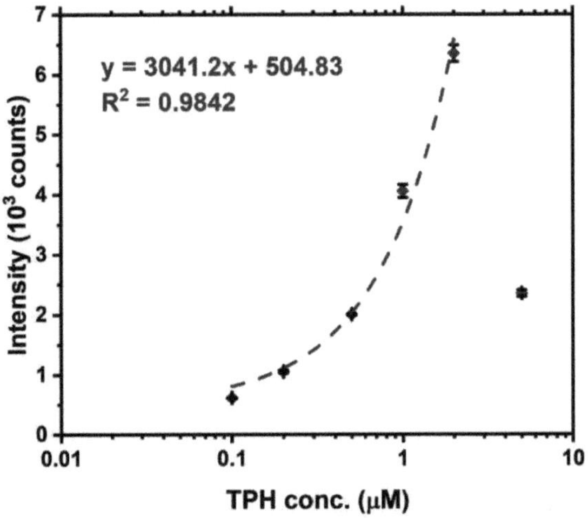

Figure 7.13 (a) Schematic illustration of the precise plasmonic hotspots within the CB8-Au NP nanosensor for TPH detection (not to scale). (b) SERS spectra of TPH with or without CB8. (c) Full-range and (d) zoom-in SERS spectra of TPH with concentrations from 0 to 2 μM. Main Raman peak of TPH at 573 cm^{-1} is marked by x. (e) Corresponding plot of SERS intensity of the main TPH peak in (d) against its concentration.

Table 7.2 The detection limit of MeX using SERS sensors

Type of SERS sensor	Detection limit of MeX	Ref.
Ag NPs	CAF (1 μM), TBR (1 μM), PRX (1 μM)	[4]
Ag NPs	CAF (1.4 μM), TBR (0.28 μM), TPH (1.4 μM)	[11]
Ag NPs	CAF (250 μM)	[12]
Ag NPs	CAF (5700 μM)	[13]
Ag NPs	CAF (1000 μM)	[9, 14]
CB7-Au NPs	TBR (0.5 μM), TPH (0.05 μM), CAF (5 μM)	[5]
CB8-Au NPs	TBR (0.05 μM), TPH (0.1 μM), CAF (1 μM)	[5]

PRX = paraxanthine (1,7-dimethylxanthine).

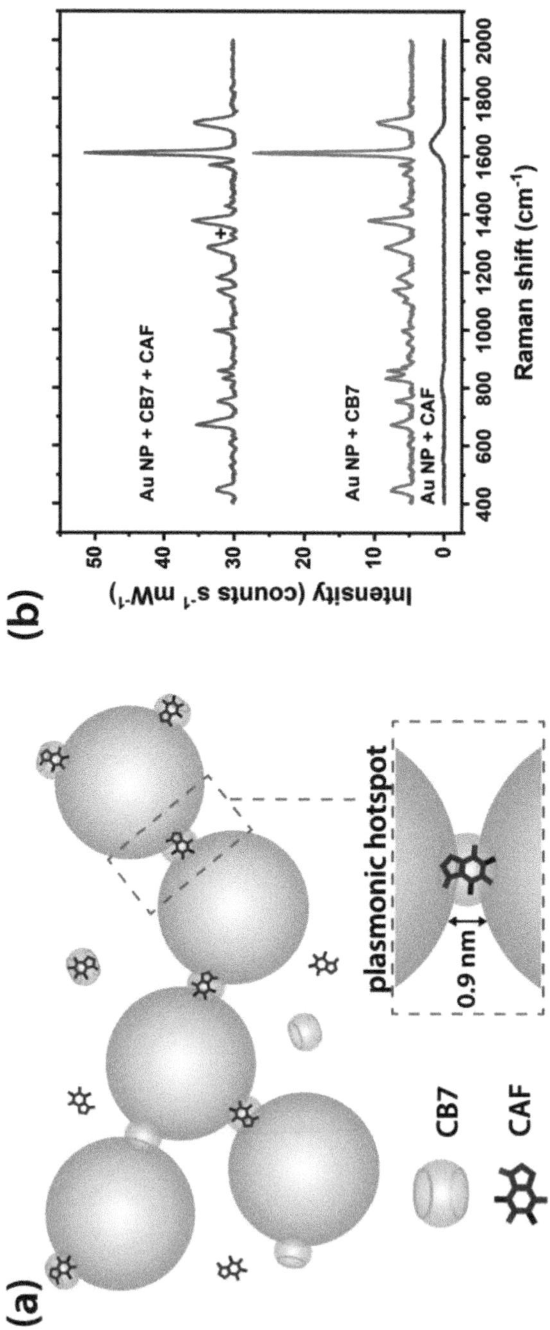
(a)
CB7
CAF
plasmonic hotspot
0.9 nm
(b)
Au NP + CB7 + CAF
Au NP + CB7
Au NP + CAF
Raman shift (cm-1)
Intensity (counts s-1 mW-1)

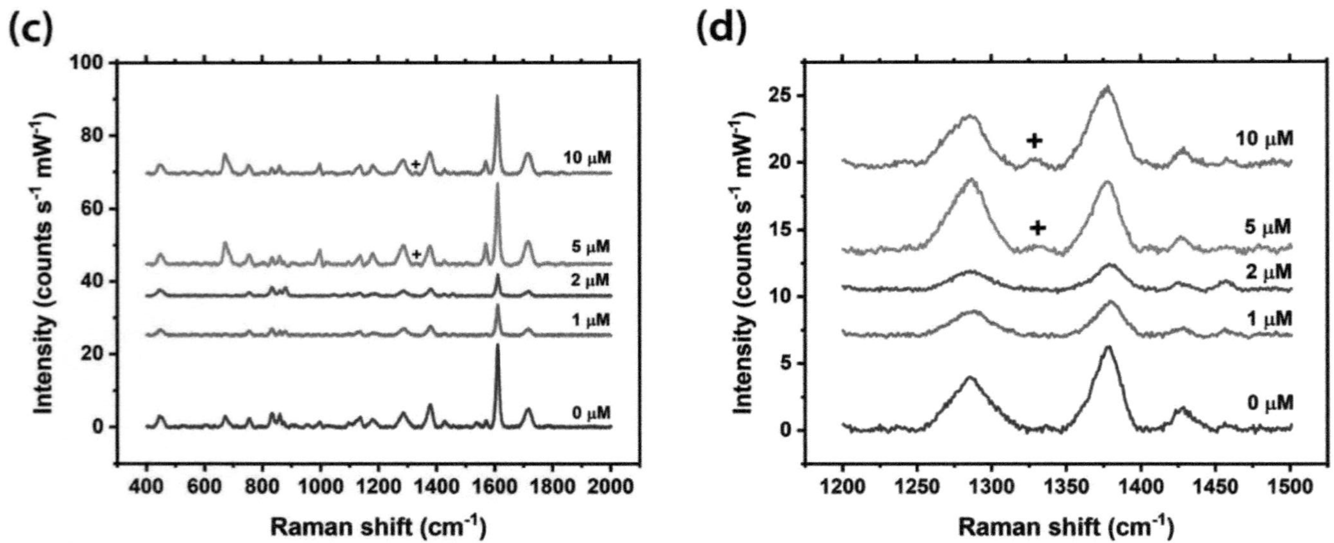

Figure 7.14 (a) Schematic illustration of the precise plasmonic hotspots within the CB7-Au NP nanosensor for CAF detection (not to scale). (b) SERS spectra of CAF with or without CB7. (c) Full-range and (d) zoom-in SERS spectra of CAF with concentrations from 0 to 10 μM. Main Raman peak of CAF at 1330 cm^{-1} is marked by +.

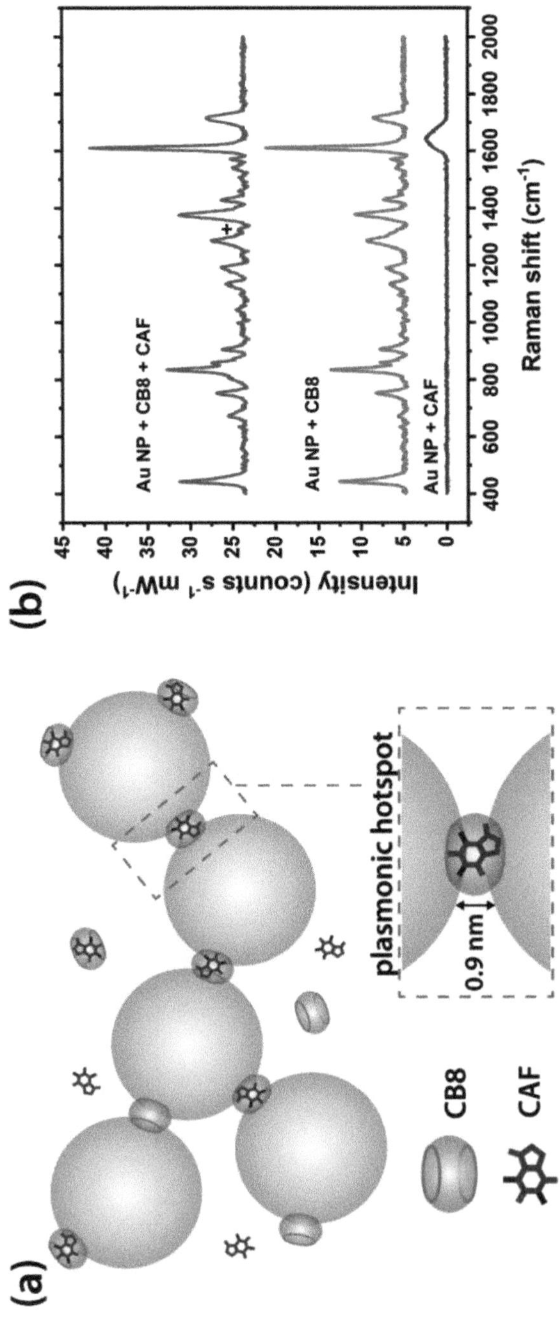
(a)
CB8
CAF
plasmonic hotspot
0.9 nm
(b)
Au NP + CB8 + CAF
Au NP + CB8
Au NP + CAF
Intensity (counts s^{-1} mW^{-1})
Raman shift (cm^{-1})
0
5
10
15
20
25
30
35
40
45
400
600
800
1000
1200
1400
1600
1800
2000

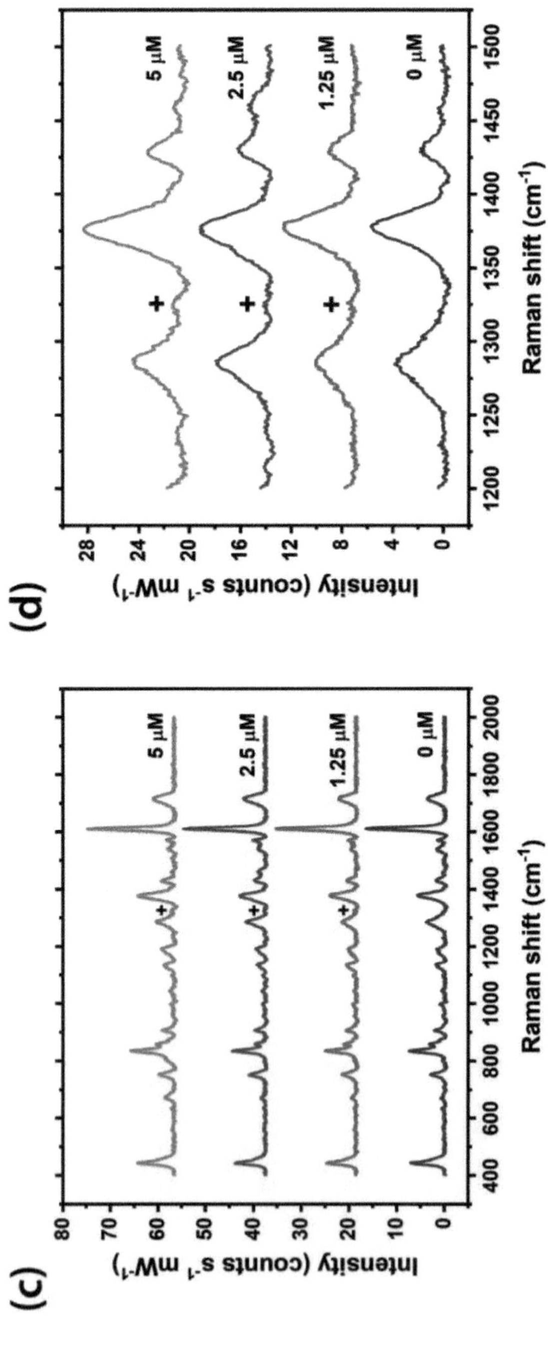
(c)
Intensity (counts s-1 mW-1)
Raman shift (cm-1)
5 μM
2.5 μM
1.25 μM
0 μM
(d)
Intensity (counts s-1 mW-1)
Raman shift (cm-1)
5 μM
2.5 μM
1.25 μM
0 μM

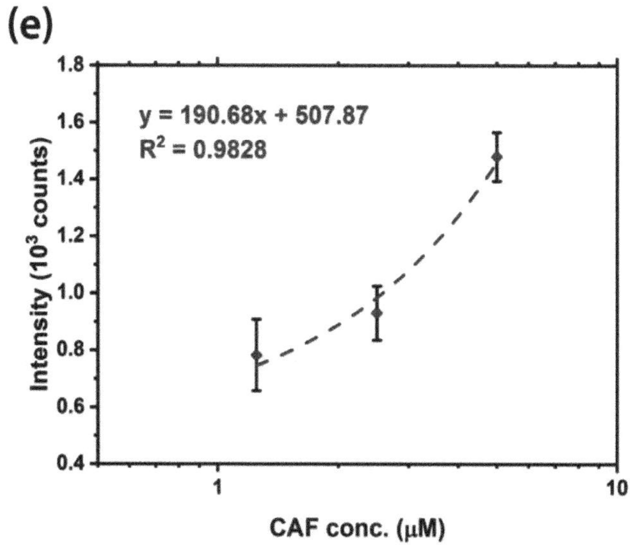

Figure 7.15 (a) Schematic illustration of the precise plasmonic hotspots within the CB8-Au NP nanosensor for CAF detection (not to scale). (b) SERS spectra of CAF with or without CB8. (c) Full-range and (d) zoom-in SERS spectra of CAF with concentrations from 0 to 5 μM. Main Raman peak of CAF at 1330 cm^{-1} is marked by +. (e) Corresponding plot of SERS intensity of the main CAF peak in (d) against its concentration.

Furthermore, CAF, the trimethylated analogue of TBR and TPH, can also be detected by the CB-Au NP nanosensor but with weaker SERS signals due to its bulkier size (Figs. 7.14 and 7.15). CAF can be detected down to 5 μM and 1.25 μM using CB7-Au NP and CB8-Au NP nanosensors, respectively, in which the characteristic Raman peak of CAF at 1326 cm^{-1} can be observed (Figs. 7.14c,d and 7.15c,d). There is a good correlation ($R^2 \sim 0.98$) between the SERS intensity and the concentration of CAF using the CB8-Au NP nanosensor (Fig. 7.15e).

7.2.4 Multiplexed SERS Sensing of Drug Isomers

The drug isomers, TBR and TPH, can be simultaneously detected by the CB8-Au NP nanosensor at different concentrations, as evidenced by the characteristic main peaks of TPH and TBR at 573 cm^{-1} and 1312 cm^{-1}, respectively, in the SERS spectra (Fig. 7.16a,b). There are good linear correlations ($R^2 \sim 0.87 - > 0.99$) between the SERS

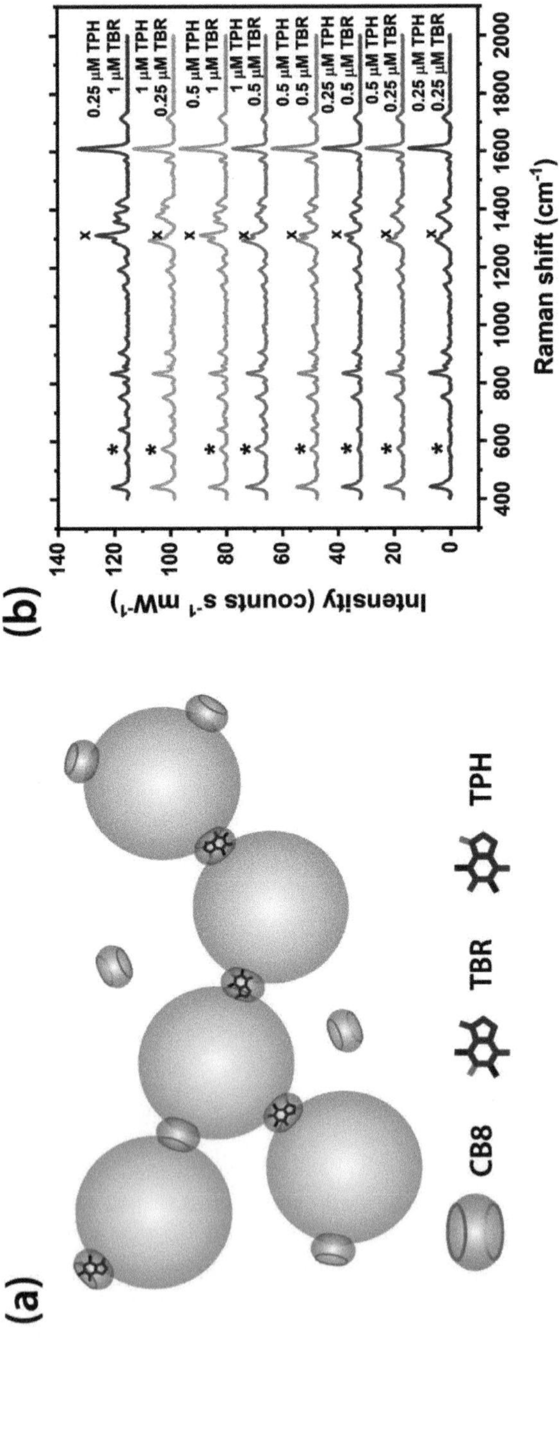
(a)
CB8
TBR
TPH
(b)
0.25 μM TPH
1 μM TBR
1 μM TPH
0.25 μM TBR
0.5 μM TPH
1 μM TBR
1 μM TPH
0.5 μM TBR
0.5 μM TPH
0.5 μM TBR
0.25 μM TPH
0.5 μM TBR
0.5 μM TPH
0.25 μM TBR
0.25 μM TPH
0.25 μM TBR
Intensity (counts s^{-1} mW^{-1})
Raman shift (cm^{-1})

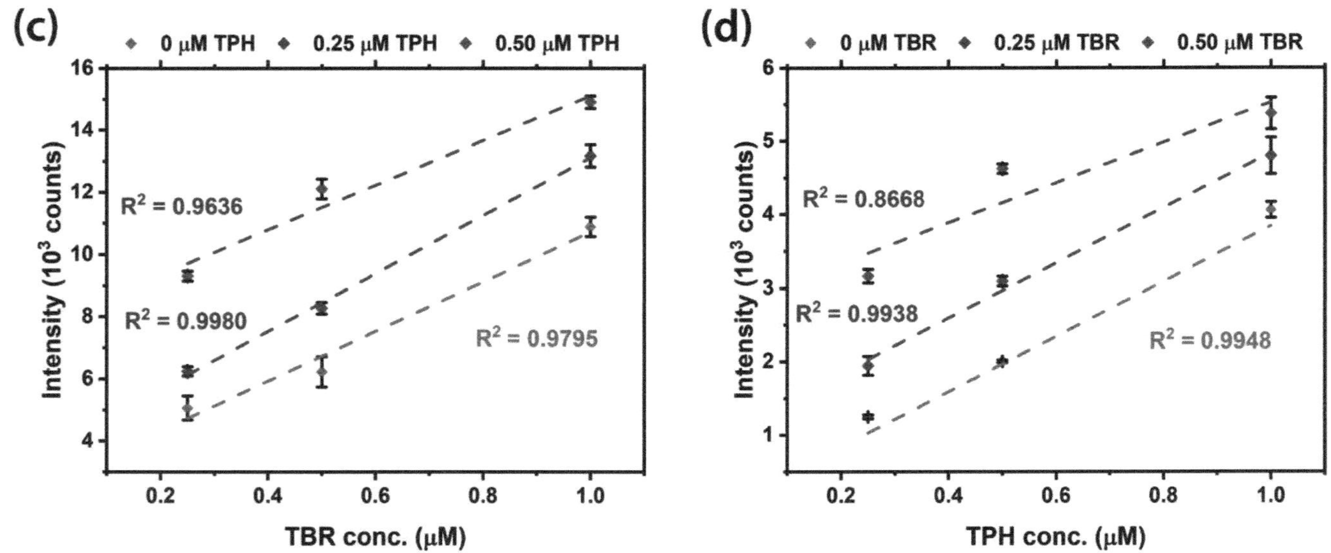

Figure 7.16 (a) Schematic illustration of the multiplexed detection of TBR and TPH using CB8-Au NP nanosensor (not to scale). (b) SERS spectra of different mixtures of TBR and TPH. Main Raman peaks of TBR and TPH are marked by x and *, respectively. (c) Corresponding plot of SERS intensity of the main TBR peak (marked by x in (b)) against its concentration with various concentrations of TPH. (d) Corresponding plot of SERS intensity of the main TPH peak (marked by * in (b)) against its concentration with various concentrations of TBR.

intensity and concentration of TBR and TPH in all mixtures with a combined concentration between 0.25 and 1 μM (Fig. 7.16c,d), thus illustrating the capability of the nanosensor for multiplexed detection of structurally similar compounds via molecular fingerprinting. Note that the detection limit of the CB8-Au NP nanosensor is even lower than previous SERS nanosensors based on Ag NPs [4, 11]. Furthermore, the small error bars (~1–18% error) show the high signal reproducibility of the nanosensor (Figs. 7.10–7.13 and 7.16).

7.2.5 Multiplexed Quantification Using Machine Learning Techniques

The detection range for analyte molecules can be extended using machine learning techniques to further analyse the dataset and tackle the nonlinear correlations at high analyte concentrations. Note that the dataset consists of all SERS spectra of TBR and TPH measured using CB8-Au NP nanosensors, including both the single analyte samples and binary mixtures. Two algorithms, partial least squared regression (PLSR) and artificial neural networks (ANNs), can be applied to predict the concentrations of TBR and TPH. The SERS spectra in the dataset can be truncated to eliminate the noisy regions near the edge of the spectral detection band of the Raman spectrometer [15]. For instance, the Raman shifts outside the region of 500–1800 cm^{-1} are removed. Then, asymmetric least squares (ALS) baseline correction is applied to the spectra before performing standard normal variate normalisation to give each spectrum a mean intensity of 0 and a standard deviation of 1 [15, 16]. For the ANN models, the analyte concentrations can be scaled into the range [0, 1] via min-max normalisation.

PLSR is a commonly used multivariate approach in which the dimensionality of the predictor variables (**X**) can be reduced by finding new latent variables to describe the variation in the response variables (**Y**) [17, 18]. In particular, the dataset is split into a matrix of **X** ($m \times n$) and a matrix of **Y** ($p \times n$) in the analysis, where m is the number of Raman shifts in the SERS spectra after pre-processing ($m = 1167$, equally spaced in the region of 500–1800 cm^{-1}), n is the number of spectra, and p is the number of analyte molecules ($p = 2$, TBR and TPH) [18, 19].

The **X** and **Y** matrices can be decomposed according to [17, 18]:

$$\mathbf{X} = \mathbf{TP}^t + \mathbf{E} \tag{7.1}$$

$$\mathbf{Y} = \mathbf{UQ}^t + \mathbf{F} \tag{7.2}$$

where **T** and **U** are the **X** and **Y** scores, **P** and **Q** are the **X** and **Y** loadings ($\mathbf{P}^t$ and $\mathbf{Q}^t$ are transpose), and **E** and **F** are the residuals.

The decomposition is performed in a way to maximise the covariance of **T** and **U**. The scores are related by:

$$\mathbf{U} = \mathbf{BT} \tag{7.3}$$

where **B** is a matrix of the PLSR coefficients.

The Python class 'sklearn.cross_decomposition.PLSRegression', which utilises the NIPALS algorithm [17], can be used to perform PLSR on pre-processed SERS spectra. The number of PLSR components used should be chosen to minimise the mean squared error.

ANNs are a biologically inspired subset of machine learning which mimic the ability of the human brain to recognise patterns and classify data. There are many interconnected nodes called artificial neurons that can take in multiple inputs to generate a single output [20]. Multilayer perceptrons (MLPs) or feedforward neural networks, which consist of an input layer, one or more hidden layers, and an output layer, are one of the most popular and simplest types of neural networks. In particular, each neuron in a particular layer connects to all neurons in the following layer, and random weights are initially assigned to the connections between neurons [15, 21]. When the network is presented with a training dataset, backpropagation is used to iteratively adjust the weights to reduce the difference between output and actual results [20, 21]. The process is repeated until the error reaches an acceptable level so the trained network can be used to determine the output for a new and unseen dataset. The Python class 'sklearn.neural_network.MLPRegressor' can be used to build an MLP and predict the concentration of analyte molecules.

The PLSR and ANN models can be trained and tested via the bootstrap resampling procedure, in which the training set is created by random selection of *n* observations from the dataset with replacement (some observations can appear more than once) [16]. Note that the number of selections is equal to the number of

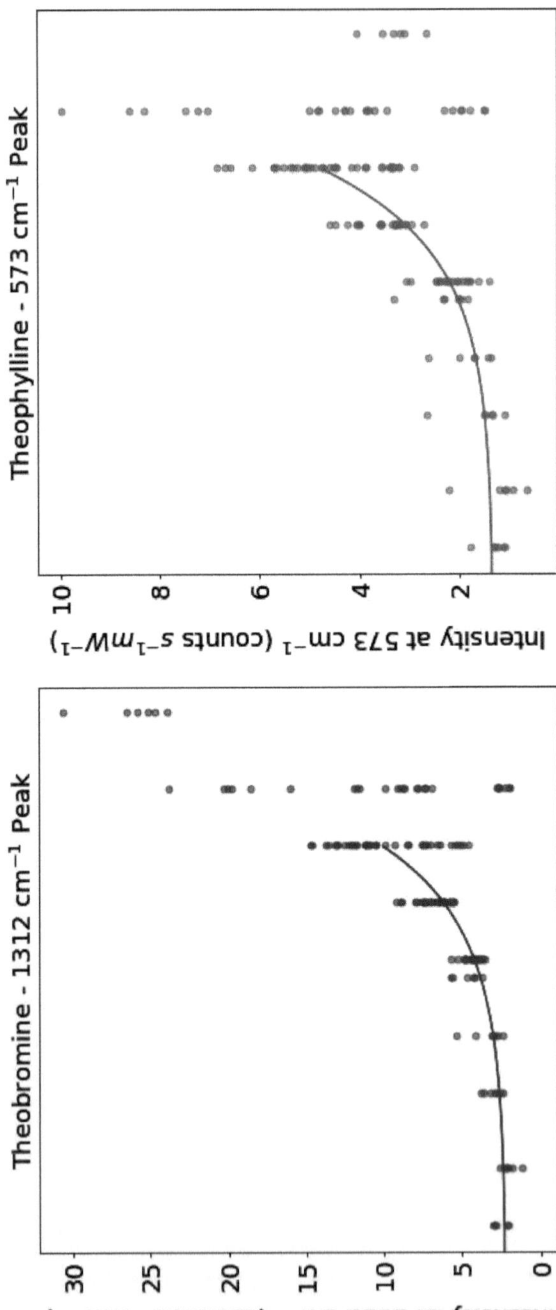

Figure 7.17 SERS intensities of the main TBR (blue) and TPH (red) peaks against their concentrations in the CB8-Au NP dataset.

observations in the original dataset. On the other side, the test set is made up of any observation that has never been selected for the training set. The mutually exclusive training and test set are used to build and evaluate the models, respectively. A total of 1000 bootstrapping iterations are performed while the mean R^2 and root mean square error of prediction (RMSEP) values are then calculated to evaluate the models. Note that the RMSEP values can also be interpreted as the detection limit.

In the binary mixtures, the intensity of the SERS peaks increases less significantly, or even decreases at higher concentrations of TBR and TPH (above 1 μM), due to disruption in the aggregation of Au NPs (Fig. 7.17). In the CB8-TBR and CB8-TPH host–guest complexes, one of the CB8 portals may become partially hindered by TBR or TPH, thus weakening the ability of the macrocycle to form plasmonic nanojunctions between the adjacent Au NPs. This implies there may not be enough free CB8 molecules to trigger the fast aggregation kinetics of the Au NPs, resulting in weaker SERS signals and absence of a linear relationship between the SERS intensity and the concentration of TBR or TPH.

The ability of the machine learning algorithms to model both linear and nonlinear relationships can be evaluated by splitting the dataset into two versions, one consists of samples with individual analyte concentrations ≤1 μM while the other contains all sample solutions with concentrations ≤5 μM. The PLSR and ANN models are trained and tested using both versions of the dataset.

Table 7.3 Optimised parameters for multilayer perceptrons

Parameter	Value
No. of input nodes	1167
No. of output nodes	2
No. of nodes in the hidden layers (concentration ≤1 μM)	16–128
No. of nodes in the hidden layers (concentration ≤5 μM)	32-32-128-16
Activation function	Rectified Linear Function (ReLU)
Algorithm for weight optimisation	Limited-memory BFGS
Maximum no. of iterations	200

After optimisation, the number of components used in the PLSR models for the linear (≤1 μM) and nonlinear (≤5 μM) datasets are 6 and 8, respectively, while the number of nodes used in the ANN models are 1167-16-128-2 and 1167-32-32-128-16-2, respectively (Table 7.3). For nonlinear datasets, more components or layers are required due to the relatively more complex relationship between the SERS intensity and the concentration of molecules.

The PLSR model performs well ($R^2 = 0.939$) for the linear dataset but shows a relatively poorer performance ($R^2 = 0.858$) for the nonlinear dataset (Figs. 7.18a and 7.19a, Table 7.4). This is due to the fact that the PLSR model is based on linear regression. In contrast, the ANN model shows better performance ($R^2 = 0.953$ for ≤1 μM, $R^2 = 0.948$ for ≤5 μM) for both datasets (Figs. 7.18b and 7.19b). The detection limits of TBR and TPH in the ANN model trained with the larger dataset are consistent with those determined directly by experiments (Figs. 7.11 and 7.13). Note that the ANN model can extend the concentration range for multiplexed detection of TBR and TPH from 1 μM to 5 μM without compromising the prediction accuracy.

Table 7.4 Mean R^2 and RMSEP values for the PLSR and ANN models

		TBR		TPH	
Detection range		R^2	**RMSEP (μM)**	R^2	**RMSEP (μM)**
0–1 μM	PLSR	0.952	0.082	0.926	0.102
	ANN	0.968	0.067	0.939	0.092
0–5 μM	PLSR	0.840	0.396	0.875	0.345
	ANN	0.962	0.193	0.934	0.252

The RMSEP values can be interpreted as the detection limit.

Furthermore, modified SERS spectra, in which the characteristic peaks of TBR and TPH are removed, can also be used to build the PLSR and ANN models (Fig. 7.20). For instance, the SERS intensities of the regions around the characteristic peaks of TBR and TPH (642 cm^{-1} and 1312 cm^{-1} for TBR and 573 cm^{-1} and 1298 cm^{-1} for TPH) are changed to zero. Note that the modified PLSR and ANN models perform only slightly worse than those trained with the original spectra (Fig. 7.21). This implies that machine learning algorithms are able to extract subtle information from datasets and justify the use of entire spectra to make robust predictions.

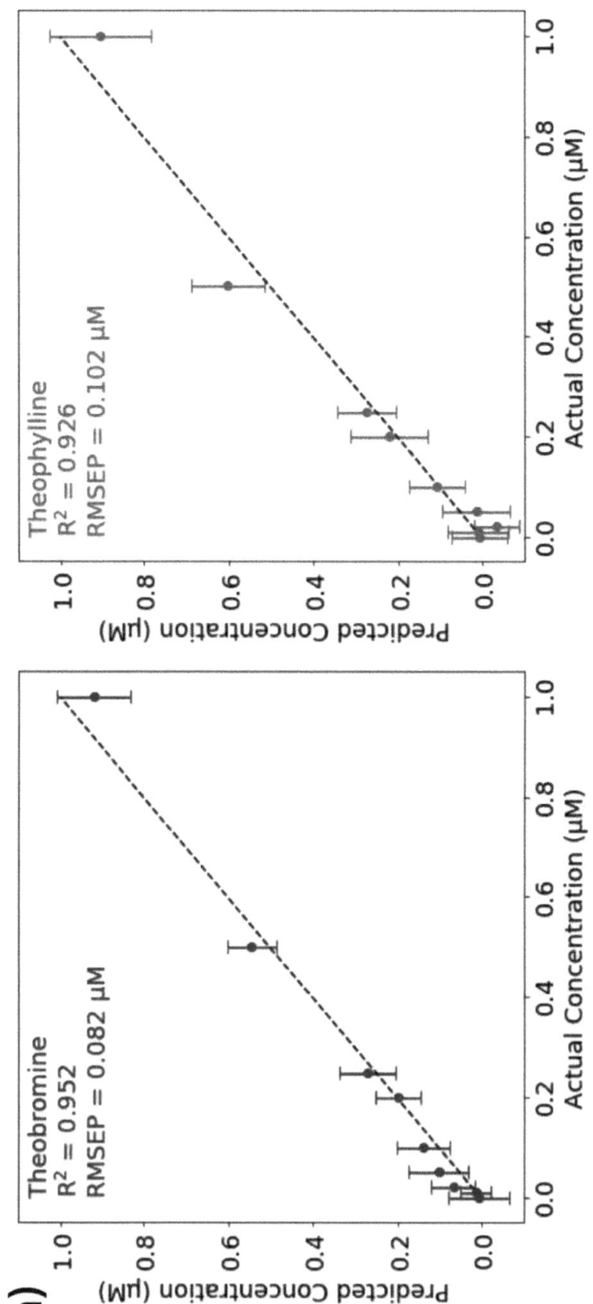
(a)
Theobromine
R^2 = 0.952
RMSEP = 0.082 µM
Theophylline
R^2 = 0.926
RMSEP = 0.102 µM
Predicted Concentration (µM)
Actual Concentration (µM)

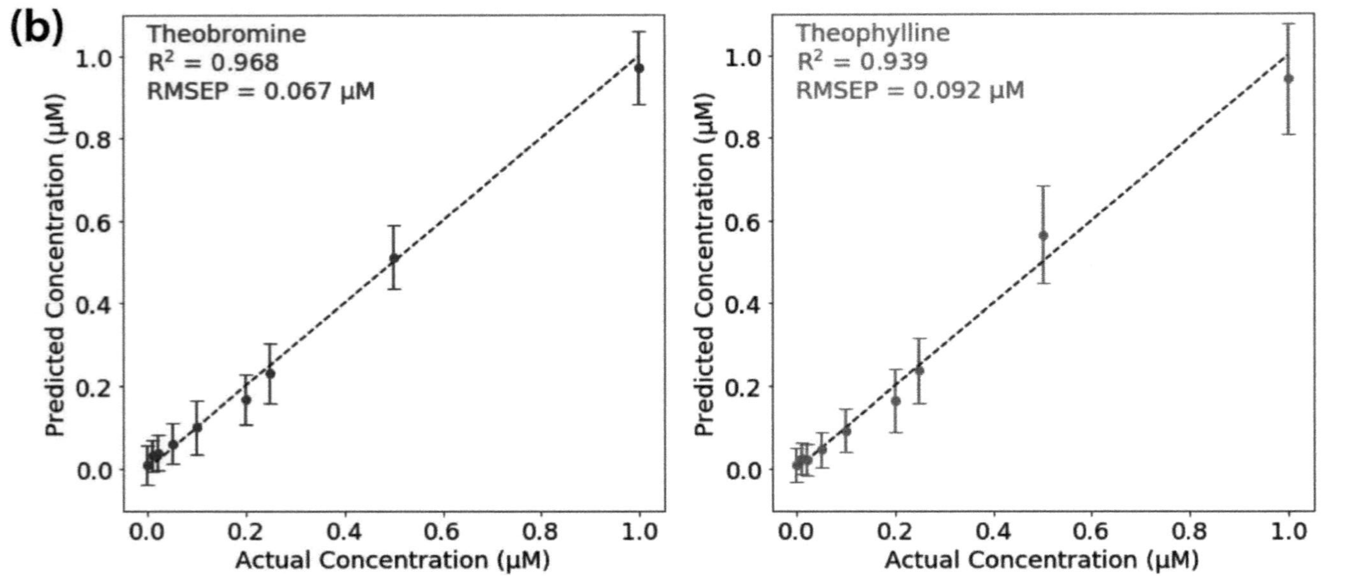

Figure 7.18 Predicted concentrations of TBR (blue) and TPH (red) made using (a) the PLSR model and (b) the ANN model trained with SERS spectra of ≤1 µM TBR and TPH from the CB8-Au NP dataset.

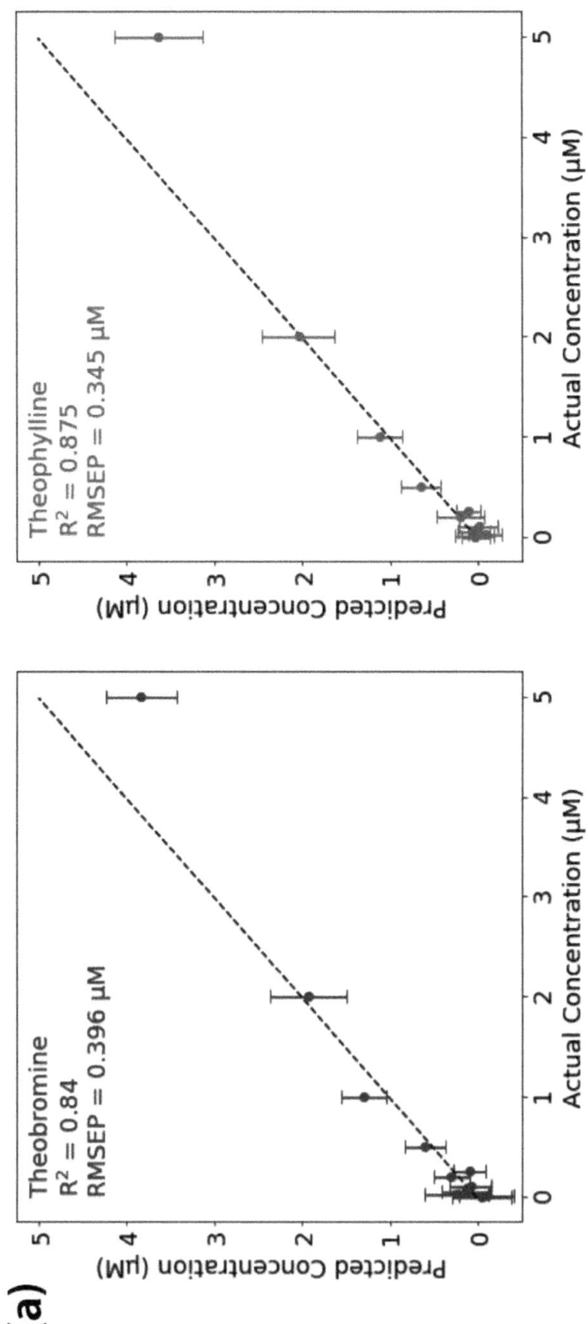
(a)
Theobromine
R² = 0.84
RMSEP = 0.396 µM
Theophylline
R² = 0.875
RMSEP = 0.345 µM
Predicted Concentration (µM)
Actual Concentration (µM)

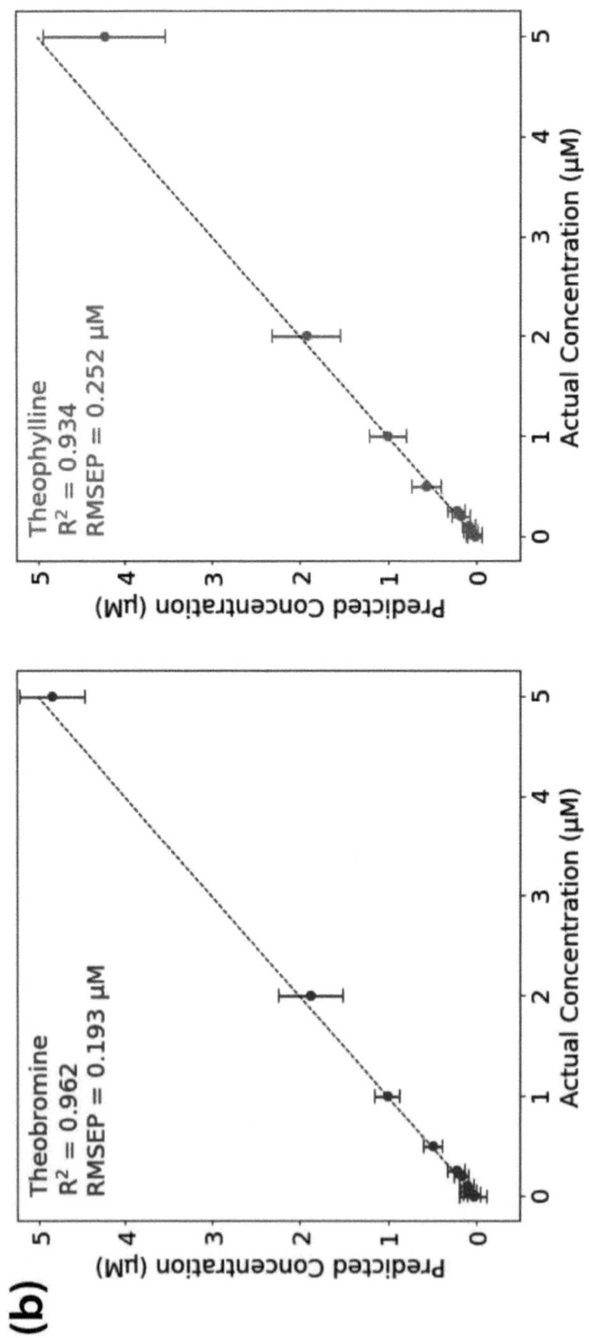

Figure 7.19 Predicted concentrations of TBR (blue) and TPH (red) made using (a) the PLSR model and (b) the ANN model trained with SERS spectra of ≤5 µM TBR and TPH from the CB8-Au NP dataset.

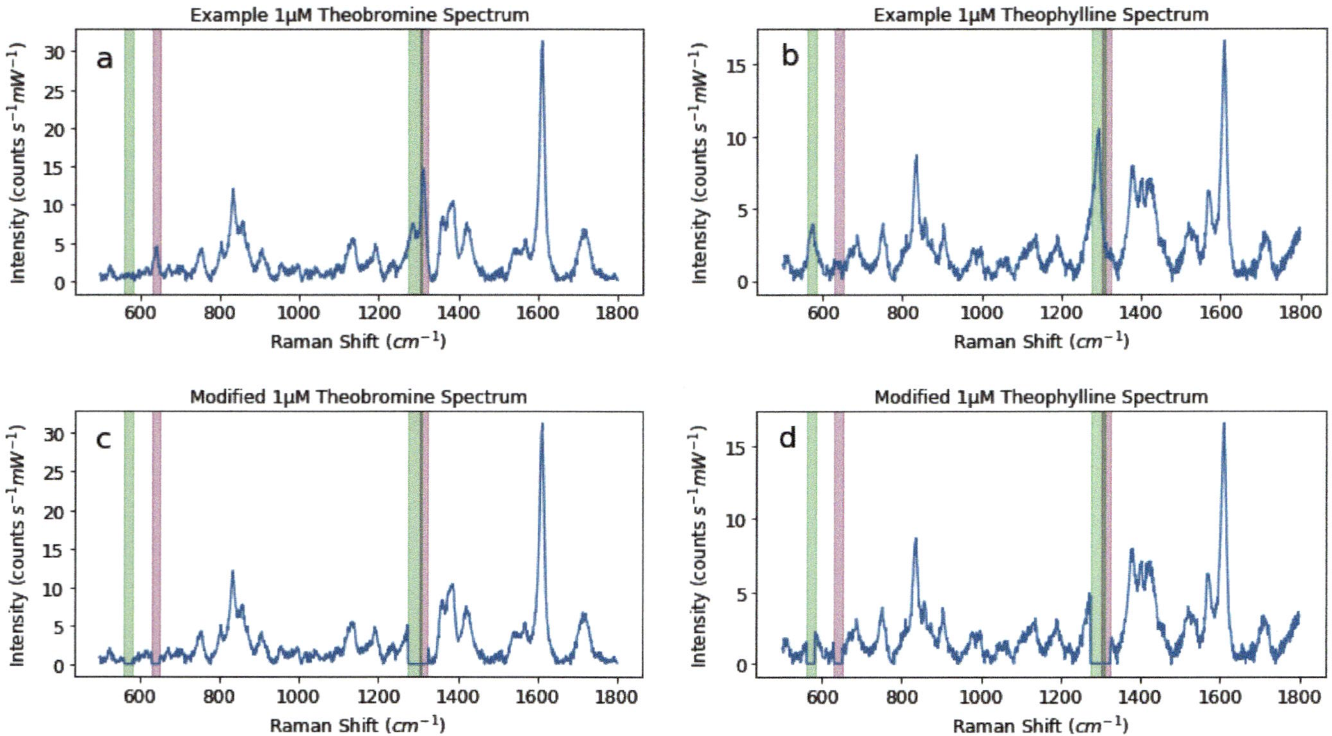

Figure 7.20 Representative SERS spectra of 1 µM TBR and 1 µM TPH from the CB8-Au NP dataset (a, b) before and (c, d) after the removal of the characteristic peaks. The relevant regions are highlighted in purple and green for TBR and TPH, respectively.

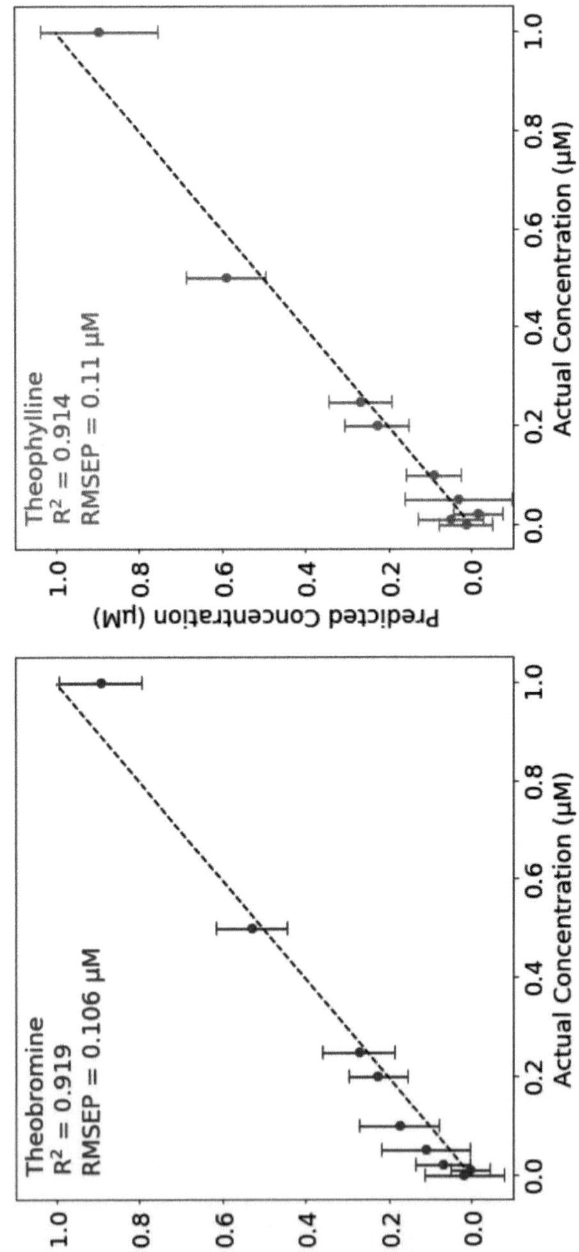
(a)
Theobromine
R² = 0.919
RMSEP = 0.106 µM
Predicted Concentration (µM)
Actual Concentration (µM)
Theophylline
R² = 0.914
RMSEP = 0.11 µM
Predicted Concentration (µM)
Actual Concentration (µM)

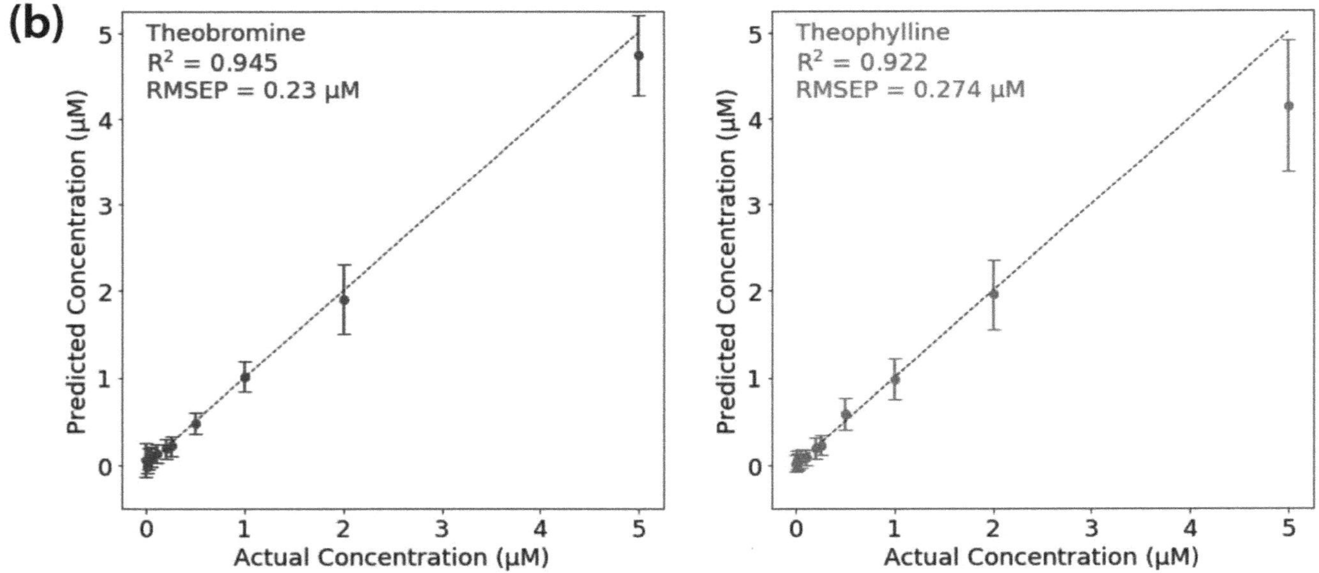

Figure 7.21 Predicted concentrations of TBR (blue) and TPH (red) made using (a) the PLSR model and (b) the ANN model trained with the modified SERS spectra of ≤1 µM and ≤5 µM TBR and TPH from the CB8-Au NP dataset, respectively.

7.3 Summary

- The CB7-Au NP and CB8-Au NP nanosensors can detect and quantify structurally similar drug molecules MeX, including TBR, TPH, and CAF, in aqueous solution via formation of host–guest complexes and precise plasmonic hotspots within the nanoaggregates.
- The key binding parameters of different CB-MeX complexes can be quantified using NMR and UV-Vis and supported by DFT molecular models.
- The binding constants of CB-MeX can be significantly enhanced by matching the molecular size of the host–guest pair.
- The detection limits of TBR and TPH using CB-Au NP nanosensors have reached ~50 nM with highly reproducible SERS signals, in contrast to the weaker signals of the bulkier CAF (1.25 μM).
- The nanosensors can simultaneously quantify multiple structurally similar molecules.
- The dynamic range of the nanosensors can be extended using machine learning algorithms including PLSR and ANNs.

References

1. Martinez-Pinilla, E., Onatibia-Astibia, A. and Franco, R. (2015). The relevance of theobromine for the beneficial effects of cocoa consumption, *Front. Pharmacol.*, **6**, 30.
2. Sanphui, P. and Nangia, A. (2014). Salts and co-crystals of theobromine and their phase transformations in water, *J. Chem. Sci.*, **126**, pp. 1249–1264.
3. Sugimoto, N., Miwa, S., Hitomi, Y., Nakamura, H., Tsuchiya, H. and Yachie, A. (2014). Theobromine, the primary methylxanthine found in *Theobroma cacao*, prevents malignant glioblastoma proliferation by negatively regulating phosphodiesterase-4, extracellular signal-regulated kinase, Akt/mammalian target of rapamycin kinase, and nuclear factor-kappa B, *Nutr. Cancer*, **66**, pp. 419–423.
4. Alharbi, O., Xu, Y. and Goodacre R. (2015). Simultaneous multiplexed quantification of caffeine and its major metabolites theobromine and paraxanthine using surface-enhanced Raman scattering, *Anal. Bioanal. Chem.*, **407**, pp. 8253–8261.

5. Chio, W.-I. K., Liu, J., Jones, T., Perumal, J., Dinish, U. S., Parkin, I. P., Olivo, M. and Lee, T.-C. (2021). SERS multiplexing of methylxanthine drug isomers via host–guest size matching and machine learning, *J. Mater. Chem. C*, **9**, pp. 12624–12632.

6. Mecozzi, S. and Rebek. J. (1998). The 55% solution: a formula for molecular recognition in the liquid state, *Chem. Eur. J.*, **4**, pp. 1016–1022.

7. Edwards, H. G. M., Munshi, T. and Anstis, M. (2005). Raman spectroscopic characterisations and analytical discrimination between caffeine and demethylated analogues of pharmaceutical relevance, *Spectrochim. Acta A*, **61**, pp. 1453–1459.

8. Gunasekaran, S., Sankari, G. and Ponnusamy, S. (2005). Vibrational spectral investigation on xanthine and its derivatives–theophylline, caffeine and theobromine, *Spectrochim. Acta A*, **61**, pp. 117–127.

9. Chen, X., Gu, H., Shen, G., Dong, X. and Kang, J. (2010). Spectroscopic study of surface enhanced Raman scattering of caffeine on borohydride-reduced silver colloids, *J. Mol. Struct.*, **975**, pp. 63–68.

10. Mahajan, S., Lee, T.-C., Biedermann, F., Hugall, J. T., Baumberg, J. J. and Scherman, O. A. (2010). Raman and SERS spectroscopy of cucurbit[*n*]urils, *Phys. Chem. Chem. Phys.*, **12**, pp. 10429–10433.

11. Wu, M., Li, H., Lv, D. and Lu, F. (2017). Dynamic-SERS spectroscopy for in situ discrimination of xanthine analogues in ternary mixture, *Anal. Bioanal. Chem.*, **409**, pp. 5569–5579.

12. Kang, J., Gu, H., Zong, L., Hu, Y. and Liu, F. (2011). The pH dependent Raman spectroscopic study of caffeine, *Spectrochim. Acta A*, **78**, pp. 757–762.

13. Pavel, I., Szeghalmi, A., Moigno, D., Cinta, S. and Kiefer, W. (2003). Theoretical and pH dependent surface enhanced Raman spectroscopy study of caffeine, *Biopolymers*, **72**, pp. 25–37.

14. Sato-Berru, R., Redon, R., Vazquez-Olmos, A. and Saniger, J. M. (2009). Silver nanoparticles synthesized by direct photoreduction of metal salts. Application in surface-enhanced Raman spectroscopy, *J. Raman Spectrosc.*, **40**, pp. 376–380.

15. Kasera, S., Herrmann, L. O., del Barrio, J., Baumberg, J. J. and Scherman, O. A. (2015). Quantitative multiplexing with nano-self-assemblies in SERS, *Sci. Rep.*, **4**, 6785.

16. James, G., Witten, D., Hastie, T. and Tibshirani, R. (2013). *An Introduction to Statistical Learning* (Springer New York, NY).

17. Wold, S., Sjostrom, M. and Eriksson, L. (2001). PLS-regression: a basic tool of chemometrics, *Chemom. Intell. Lab. Syst.*, **58**, pp. 109–130.

18. Kreeger, P. K. (2013). Using partial least squares regression to analyse cellular response data, *Sci. Signal.*, **6**, p. tr7.

19. Eriksson, L., Byrne, T., Johansson, E., Trygg, J. and Vikstrom, C. (2013). *Multi- and Megavariate Data Analysis Basic Principles and Applications* (Umetrics Academy, Sweden).

20. Hallinan, J. S. (2013). *Methods in Microbiology,* **40**, eds. Harwood, C. and Wipat, A. (Academic Press, Netherlands), pp. 1–37.

21. Lussier, F., Thibault, V., Charron, B., Wallace, G. Q. and Masson, J.-F. (2020). Deep learning and artificial intelligence methods for Raman and surface-enhanced Raman scattering, *TrAC Trends Anal. Chem.*, **124**, 115796.

Index